LAB WORKBOOK

NINTH EDITION
Modern Plumbing

by **Charles H. Owenby**

Publisher
The Goodheart-Willcox Company, Inc.
Tinley Park, IL
www.g-w.com

Copyright © 2022
by
The Goodheart-Willcox Company, Inc.

All rights reserved. No part of this work may be reproduced, stored, or transmitted in any form or by any electronic or mechanical means, including information storage and retrieval systems, without the prior written permission of
The Goodheart-Willcox Company, Inc.

Manufactured in the United States of America.

ISBN 978-1-64564-670-9

3 4 5 6 7 8 9 – 22 – 25 24 23 22

The Goodheart-Willcox Company, Inc. Brand Disclaimer: Brand names, company names, and illustrations for products and services included in this text are provided for educational purposes only and do not represent or imply endorsement or recommendation by the author or the publisher.

The Goodheart-Willcox Company, Inc. Safety Notice: The reader is expressly advised to carefully read, understand, and apply all safety precautions and warnings described in this book or that might also be indicated in undertaking the activities and exercises described herein to minimize risk of personal injury or injury to others. Common sense and good judgment should also be exercised and applied to help avoid all potential hazards. The reader should always refer to the appropriate manufacturer's technical information, directions, and recommendations; then proceed with care to follow specific equipment operating instructions. The reader should understand these notices and cautions are not exhaustive.

The publisher makes no warranty or representation whatsoever, either expressed or implied, including but not limited to equipment, procedures, and applications described or referred to herein, their quality, performance, merchantability, or fitness for a particular purpose. The publisher assumes no responsibility for any changes, errors, or omissions in this book. The publisher specifically disclaims any liability whatsoever, including any direct, indirect, incidental, consequential, special, or exemplary damages resulting, in whole or in part, from the reader's use or reliance upon the information, instructions, procedures, warnings, cautions, applications, or other matter contained in this book. The publisher assumes no responsibility for the activities of the reader.

The Goodheart-Willcox Company, Inc. Internet Disclaimer: The Internet resources and listings in this Goodheart-Willcox Publisher product are provided solely as a convenience to you. These resources and listings were reviewed at the time of publication to provide you with accurate, safe, and appropriate information. Goodheart-Willcox Publisher has no control over the referenced websites and, due to the dynamic nature of the Internet, is not responsible or liable for the content, products, or performance of links to other websites or resources. Goodheart-Willcox Publisher makes no representation, either expressed or implied, regarding the content of these websites, and such references do not constitute an endorsement or recommendation of the information or content presented. It is your responsibility to take all protective measures to guard against inappropriate content, viruses, or other destructive elements.

Cover image credits: Marcel Derweduwen/Shutterstock.com (inset); OlegDoroshin/Shutterstock.com

Introduction

This lab workbook is designed to be used with the ***Modern Plumbing*** textbook. The Chapter Review questions and the Jobs are aimed at helping you master the subject matter provided in the textbook. The questions will help you remember important ideas, theories, and concepts. The Jobs will incorporate these ideas to help you improve your hands-on techniques and skills. Each Chapter Review corresponds to a chapter in the textbook, while a Job may incorporate the ideas from several chapters of the textbook.

After studying a chapter in the textbook, complete the Test Your Knowledge questions and the Suggested Activities. Once these are finished, complete the corresponding Chapter Review. After studying each chapter and successfully completing the Chapter Review, complete the Job(s) that relate to the chapter(s) you are studying.

Chapter Reviews

Chapter Reviews are to be used after studying a chapter in the textbook. Answer as many of the questions as possible without using your textbook. Complete the remaining questions by referring to the textbook where the topic is discussed. These Chapter Reviews also can be used as pretests to determine your knowledge level prior to studying a chapter.

Jobs

Jobs are to be completed after studying the textbook chapter(s) and the corresponding Chapter Review(s). The Jobs allow you to apply the ideas, theories, and concepts learned in the textbook. The Jobs will incorporate these ideas to help you improve your hands-on techniques and skills.

Jobs may consist of a Text Reference, Objective, Introduction, Tools and Equipment section, and Instructions. The Text Reference lists the textbook chapter(s) that is to be used as reference for the Job. The Objective details the purpose and the goal of the Job. The Introduction contains general information about the need and purpose of the Job. The Tools and Equipment section is a detailed list of the tools and equipment needed to complete the Job. Gather all the tools and equipment on the list before starting the Job. The Instructions are a numbered, step-by-step procedure for completing the Job. Each step is followed by a "Completed" checkbox. Upon completion of each step, place a check in the box. Throughout the Jobs, there are highlighted notes that give you additional information about the Job or a step. There are also highlighted Cautions and Warnings that give you additional safety information.

After completing the Job, have your instructor evaluate your work and initial and date the Job sheet. You will be evaluated on the final product, following the procedural steps, tool and equipment selection, return of tools and equipment, cleanliness of work area, and general shop safety. *Although time is a factor,* **safety** *and* **work quality** *are not to be sacrificed for speed.*

Safety

Safety is the number one priority when participating in hands-on jobs or when doing any plumbing work. The work area, or shop, is a place to work. It is not a place for "horseplay" or joke playing.

Your work area should always be clean and clear of debris. Make sure the proper tools and equipment are selected to complete the job. When your work is complete, make sure that all tools and equipment are in working order and are placed back in their proper places. Make sure the work area is clean when you have completed the job.

Always protect yourself. Wear safety glasses, goggles, or a face shield when working in the shop. Dress properly. Wear protective gloves, and avoid wearing loose-fitting clothing. Long sleeves should be rolled up, and if you wear a tie, remove it. Sturdy shoes with thick soles should be worn. Shoes with leather soles should be avoided. Leather soles tend to have less traction. Rings, watches, and other jewelry should always be removed.

Always handle sharp and/or pointed objects with care. Tools such as scribers, cutting tools, and screwdrivers should be placed down with the cutting edges or point facing away from you. Avoid carrying any tool in your pocket.

When using solvents and adhesives, always read the instructions carefully. Some of these fluids require a well-ventilated work area because of the fumes that they emit. Some can also cause skin irritation if they come into contact with your skin. These fluids may also be flammable. Approved fire prevention practices should be followed in the work area while working with such fluids.

When using a torch or a melting furnace, substantial heat can be produced. Care must be taken when handling heated materials. When handling or moving with heated materials, make sure your walking path is clear. If you are handling a heated liquid and it is to be set down, make sure it is set down on a level surface.

Knowledge of first aid is very important. You should know and understand common first-aid procedures. You should know where to locate the first-aid kit in the shop.

Your safety and the safety of those around you is the highest priority. Think first before you perform any Job or procedural step. If you are not sure about how to perform the procedure, ask your instructor for further assistance.

The Jobs are intended to give you the opportunity to improve your hands-on skills as they pertain to the plumbing field. Do not attempt these Jobs unless you are under the supervision of a trained plumbing instructor.

Table of Contents

Chapter Review

CHAPTER 1	Safety	1
CHAPTER 2	Plumbing Tools	7
CHAPTER 3	Plumbing Career Opportunities	11
CHAPTER 4	Plumbing History	15
CHAPTER 5	Leveling Instruments	19
CHAPTER 6	Mathematics for Plumbers	21
CHAPTER 7	Hydraulics and Pneumatics	27
CHAPTER 8	Print Reading and Sketching	29
CHAPTER 9	Green Construction	33
CHAPTER 10	Building and Plumbing Codes	35
CHAPTER 11	Soldering, Brazing, and Welding	37
CHAPTER 12	Excavating	39
CHAPTER 13	Water Supply Systems	41
CHAPTER 14	Water Treatment	45
CHAPTER 15	Plumbing Fixtures	49
CHAPTER 16	Piping Materials and Fittings	55
CHAPTER 17	Valves and Meters	59
CHAPTER 18	Water Heaters	63
CHAPTER 19	Designing Plumbing Systems	69
CHAPTER 20	Preparing for Plumbing System Installation	73
CHAPTER 21	DWV Pipe and Fitting Installation	75
CHAPTER 22	Installing Water Supply Piping	79
CHAPTER 23	Supporting and Testing Pipe	83
CHAPTER 24	Installing Water Heaters, Fixtures, Faucets, and Appliances	87
CHAPTER 25	Private Septic Systems	91
CHAPTER 26	Storm Water and Sumps	95
CHAPTER 27	Installing HVAC Systems	97
CHAPTER 28	Swimming Pools, Hot Tubs, and Spas	101
CHAPTER 29	Fire Safety and Irrigation Systems	103
CHAPTER 30	Repairing DWV Systems	107
CHAPTER 31	Repairing Water Supply Systems	111
CHAPTER 32	Remodeling	117
CHAPTER 33	Job Organization	121

Jobs

JOB 1	Safety	123
JOB 2	Math Calculations	125
JOB 3	Calculating Slope on a DWV Horizontal Sewer	127
JOB 4	Print Drawing	129
JOB 5	Print Specification Interpretation	131
JOB 6	Fitting Identification	139
JOB 7	Valve Identification	145
JOB 8	DWV Pipe and Fitting Identification	147
JOB 9	DWV Fittings—No-Hub	149
JOB 10	Soldering Copper Pipe and Fittings	151
JOB 11	Joining PEX Piping Using Crimp Connections	155
JOB 12	Joining Cast-Iron No-Hub Pipe	157
JOB 13	Joining PVC Pipe	159
JOB 14	Joining Galvanized Steel Pipe and Sleeve Coupling	163
JOB 15	Cutting and Threading Galvanized Steel Pipe	165
JOB 16	Water Supply	169
JOB 17	Joining Copper Pipe Using Compression Fittings	171
JOB 18	Flaring and Connecting Copper Tubing	173
JOB 19	Faucet and Rotating Ball Faucet Identification	177
JOB 20	Installing a Toilet (Tank-Type)	179
JOB 21	Rough-In Residential Bathroom Group	181

Name _____ Date _____ Class _____

CHAPTER 1 / Safety

OBJECTIVE: You will be able to develop a list of general safety rules relating to clothing, ladders, electrical tools, and scaffolds and explain why it is necessary to develop safe working habits.

Carefully study the chapter and then answer the following questions.

1. _____ *True or False?* Safety is one aspect of plumbing that is often neglected.

2. Carelessness, lack of knowledge, substance _____, and defective tools and equipment all contribute to accidents on the job.
   _____

3. _____ Good _____ means keeping the work area as clean and orderly as possible.
   A. procedure
   B. housekeeping
   C. work
   D. cleanliness

4. _____ _____ work habits are a result of attitudes formed by the worker.
   A. Constant
   B. Sufficient
   C. Unnecessary
   D. Safe

5. _____ Dressing _____ can reduce the possibility of an accident or injury.
   A. in more PPE than needed
   B. in loose clothing
   C. appropriately
   D. in old, torn clothes

6. The National Safety Council estimates that _____% of the eye injuries that occur annually in the United States could be prevented by wearing protective eyewear.
   _____

7. _____ If you wear corrective lenses, they must meet the _____ standard for safety glasses.
   A. ANSI
   B. OSHA
   C. local
   D. HCS

8. _____ Corrective lenses worn on the job should be fitted with _____.
   A. side shields
   B. a face mask
   C. safety goggles
   D. Both A and C.

9. _____ *True or False?* Wearing pants that drag on the floor is appropriate as long as they have cuffs.

10. _____ Lifting an object incorrectly is most likely to cause serious injury to your _____.
    A. legs
    B. back
    C. arms
    D. shoulders

Copyright Goodheart-Willcox Co., Inc.
May not be reproduced or posted to a publicly accessible website.

11. _____ When lifting, keep your back straight and use your _____ muscles to raise the object.
    A. leg
    B. back
    C. arm
    D. shoulder

12. _____ *True or False?* When lifting, repeated minor strains may result in permanent damage.

13. _____ When carrying large pieces of pipe, use a _____ to make the job easier and safer.
    A. lever hoist
    B. rope
    C. forklift
    D. hoist standard

14. _____ *True or False?* Most ladder-related injuries are caused by using the wrong ladder for the job, or by misusing or abusing climbing equipment.

15. _____ _____ ladders are often used for indoor work.
    A. Straight
    B. Extension
    C. Step
    D. Articulated

16. The duty rating of the ladder is its maximum _____ load capacity.

_____

17. _____ A(n) _____ ladder should *not* be used near electrical wire because it will conduct electricity.
    A. fiber-reinforced
    B. aluminum
    C. wooden
    D. heavy-duty

18. _____ An unsecured straight or extension ladder should be _____ by a fellow worker to prevent it from slipping.
    A. equipped with safety feet
    B. held
    C. equipped with blocking
    D. raised

19. _____ When using a stepladder, never work higher than the _____ rung from the top of the ladder.
    A. first
    B. second
    C. third
    D. last

20. _____ Scaffolds 4' to 10' high and less than _____" wide must also be guarded with rails.
    A. 4
    B. 10
    C. 42
    D. 45

21. _____ *True or False?* A scaffold plank board should only be placed on the top of guardrails if it is to gain greater height.

22. _____ Corded electrical tools must be properly grounded or _____.
    A. insulated
    B. double insulated
    C. charged
    D. cleaned

23. _____ *True or False?* The letters GFCI are used to identify a ground fault circuit interrupter.

Name _____

24. _____ Never use tools with _____ cords, or loose and _____ switches.
   A. long; broken
   B. short; frayed
   C. frayed; broken
   D. multiple; frayed

25. _____ *True or False?* Compressed gas cylinders are a potential fire hazard.

Match each of the following classes of fire with its description for questions 26–30.

26. _____ Combustible metals such as sodium, magnesium, and titanium; extinguishing agents that will not react to burning metals.

27. _____ Fire in "hot" electrical equipment is especially dangerous because only nonconductive extinguishing agents can safely be used.

28. _____ Flammable liquids such as gasoline, grease, and oil preventing oxygen from mixing with the vapors of the liquids.

29. _____ Ordinary combustible materials, paper, wood, cloth, rubber.

30. _____ Kitchen fires; using dry and wet chemicals.

A. Class A fires
B. Class B fires
C. Class C fires
D. Class D fires
E. Class K fires

31. _____ Excavating and trenching work involves three major types of safety. They are protecting _____, workers in trenches, and using pneumatic and hand tools safely.
   A. equipment
   B. existing underground utilities
   C. the environment
   D. wildlife

32. _____ The two categories of fall protection are fall restraint and fall _____.
   A. assist
   B. protection
   C. prevention
   D. arrest

33. _____ Edge cutting tools must be _____ to work properly.
   A. dull
   B. sharp
   C. cleaned before every use
   D. grounded

34. _____ Mushroomed heads on chisels and similar tools must be _____ to prevent small pieces from breaking off when struck.
   A. sharpened
   B. dulled
   C. removed
   D. equipped with a guard

35. Many materials used on a construction site are _____ hazards to the worker.

_____

36. A federal standard was adopted to provide information to workers about the materials they are working with and the potential hazards these materials present. This standard is abbreviated _____.

_____

37. Employers are required to make information available to employees about both physical and _____ hazards.

    _____

38. Producers or suppliers of hazardous materials must _____ all containers of hazardous materials to identify the chemicals and provide appropriate warnings.

    _____

39. A(n) _____ contains detailed information on each chemical in a specific product.

    _____

40. Identify the following acronyms.

    A. TLV _____

    B. PEL _____

    C. TWA _____

    D. STEL _____

    E. SDS _____

41. The four ways that hazardous chemicals or materials can enter the body are through _____, ingestion, absorption, and injection.

    _____

42. A(n) _____ will protect the worker from inhaling hazardous chemicals.

    _____

43. _____ Vials of _____ oil are used to test the fit of a respirator.
    A. olive
    B. vegetable
    C. banana
    D. coconut

44. _____ Full-face respirators provide eye protection and are less likely to leak at the joint between the filter and _____.
    A. mouth
    B. face
    C. hose
    D. cartridge

45. Before using a _____, it should be inspected for damage or wear before and after each use.

    _____

46. _____ Hearing loss can result from brief exposure to very loud sounds, and from extended exposure to _____ sounds.
    A. less intense
    B. high-pitched
    C. low-pitched
    D. hazardous

47. According to NIOSH, a worker must *not* enter an area where the oxygen level is below _____%.

    _____

Name _____

Match the effect caused by an oxygen-deficient atmosphere on the body at the following levels for questions 48–51.

48. _____ Difficult breathing, death in minutes
49. _____ Impaired judgement and breathing
50. _____ Minimum for safe working
51. _____ Faulty judgement, rapid breathing

A. 19.5%
B. 16%
C. 14%
D. 6%

52. _____ When material is being brazed or soldered in a "confined space," a(n) _____ can be generated.
    A. oxygen-deficient atmosphere
    B. flammable atmosphere
    C. toxic atmosphere
    D. suitable atmosphere

53. Working safely in a confined space requires forced _____ at a minimum.

_____

54. The two most common bloodborne pathogens are hepatitis B virus (HBV) and _____.

_____

Notes

Name _____ Date _____ Class _____

CHAPTER 2
Plumbing Tools

OBJECTIVE: You will be able to identify the various plumbing tools and their uses. You will be able to select the proper tool for the desired task and explain how to maintain common plumbing tools.

Carefully study the chapter and then answer the following questions.

1. _____ Measuring tools are used to measure _____.
   A. length, height, and diameter
   B. levelness
   C. plumbness
   D. All of the above.

2. Layout tools are used to produce accurate lines, _____, or any other marking.
   _____

3. _____ *True or False?* Common lengths of long steel measuring tapes are 25', 50', 100', 200', and 300' lengths.

4. _____ Plumber's rules are special folding rules available in 6' and _____' lengths.
   _____

5. _____ Combination squares can measure _____° angles.
   A. 90
   B. 45
   C. 60
   D. Both A and B.

6. _____ The three most common marking tools used to mark various types of pipe are yellow keel, pencil, and _____.
   A. markers
   B. soapstone
   C. pen
   D. chalk

7. _____ Plumb is to vertical as level is to _____.
   A. straight
   B. slope
   C. horizontal
   D. parallel

8. _____ *True or False?* Chalk lines are used to lay out long, straight lines on hard and rather smooth surfaces.

9. _____ A(n) _____ is an instrument that has a pencil in one leg and is used to lay out circles and arcs.
   A. compass
   B. divider
   C. chalk box
   D. transit

Copyright Goodheart-Willcox Co., Inc.
May not be reproduced or posted to a publicly accessible website.

Match the toothed cutting tools with their proper names for questions 10–13.

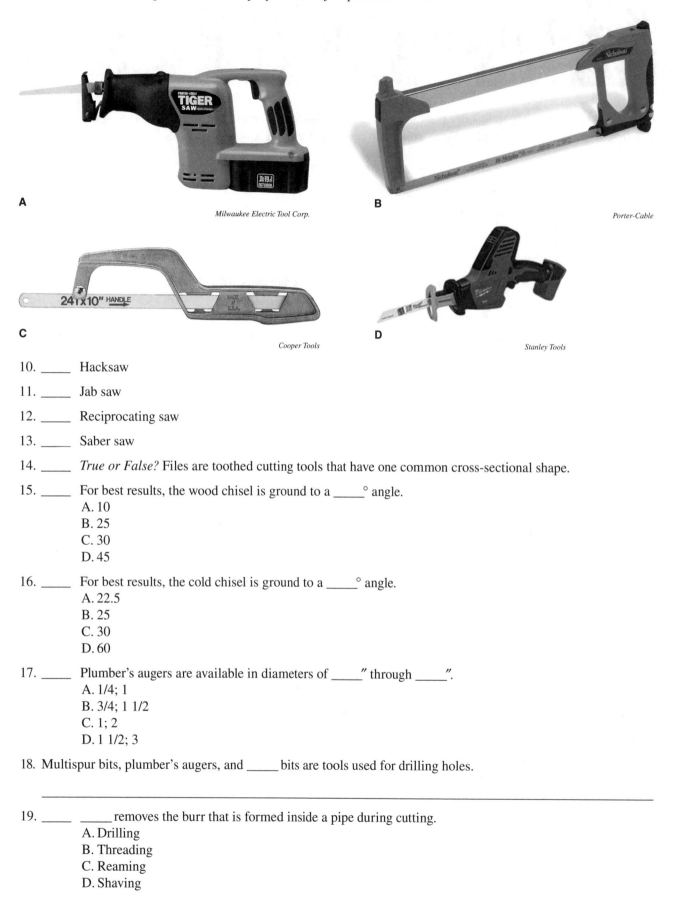

A
Milwaukee Electric Tool Corp.

B
Porter-Cable

C
Cooper Tools

D
Stanley Tools

10. _____ Hacksaw

11. _____ Jab saw

12. _____ Reciprocating saw

13. _____ Saber saw

14. _____ *True or False?* Files are toothed cutting tools that have one common cross-sectional shape.

15. _____ For best results, the wood chisel is ground to a _____° angle.
    A. 10
    B. 25
    C. 30
    D. 45

16. _____ For best results, the cold chisel is ground to a _____° angle.
    A. 22.5
    B. 25
    C. 30
    D. 60

17. _____ Plumber's augers are available in diameters of _____" through _____".
    A. 1/4; 1
    B. 3/4; 1 1/2
    C. 1; 2
    D. 1 1/2; 3

18. Multispur bits, plumber's augers, and _____ bits are tools used for drilling holes.

_____

19. _____ _____ removes the burr that is formed inside a pipe during cutting.
    A. Drilling
    B. Threading
    C. Reaming
    D. Shaving

Name _____

Match each of the following with its proper name for questions 20–24.

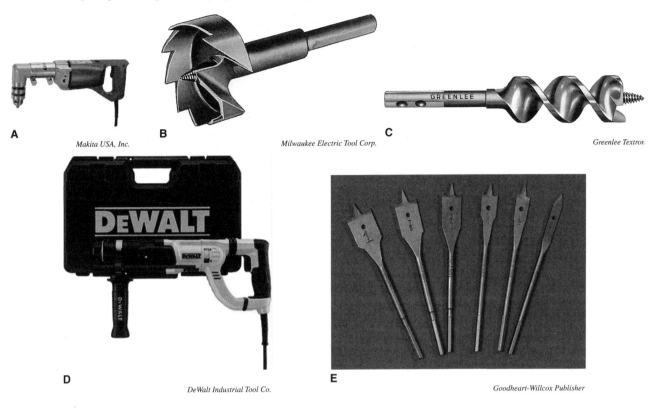

A — Makita USA, Inc.
B — Milwaukee Electric Tool Corp.
C — Greenlee Textron
D — DeWalt Industrial Tool Co.
E — Goodheart-Willcox Publisher

20. _____ Multispur bit

21. _____ Offset portable drill

22. _____ Rotary hammer drill

23. _____ Plumber's auger

24. _____ Spade bit

25. _____ A _____ threading die is a special die used for threading galvanized steel pipe.
    A. special
    B. bolt
    C. pipe
    D. straight

26. _____ _____ are sometimes needed to hold plumbing parts while the operations are performed on them.
    A. Drills
    B. Vises
    C. Pipes
    D. Reamers

27. _____ To protect pipe dies when cutting threads, _____ is applied to the threads.
    A. machine oil
    B. cutting oil
    C. 30 weight motor oil
    D. water

28. _____ *True or False?* The three-way pipe die and stock permits three diameters of pipe to be threaded with a single tool.

29. _____ A pipe wrench is used to _____ threaded pipe during assembly.
    A. hold or turn
    B. adjust
    C. ream
    D. None of the above.

Chapter 2   Plumbing Tools   **9**

30. The three basic designs of pipe wrenches are end, straight, and _____.

31. _____ A _____ wrench is used to assemble chrome-plated pipe so damage to the pipe does not occur.
    A. strap
    B. chain
    C. pipe
    D. None of the above.

Match each of the wrenches with its proper name for questions 32–36.

A

B *The Ridge Tool Co.*

The Ridge Tool Co.

C

D *The Ridge Tool Co*

E *Cooper Tools*

The Ridge Tool Co

32. _____ Pipe wrench

33. _____ Chain wrench

34. _____ Strap wrench

35. _____ Basin wrench

36. _____ Adjustable wrench

37. A(n) _____ vise is the most common holding device used by plumbers.

38. _____ A _____ hammer is used for driving or pulling nails and for tapping a wood chisel.
    A. ball peen
    B. sledge
    C. carpenter's
    D. None of the above.

39. _____ A(n) _____ screwdriver generally has two straight and two Phillips blades.
    A. 4-in-1
    B. straight
    C. Phillips
    D. None of the above.

Name _____ Date _____ Class _____

CHAPTER 3
Plumbing Career Opportunities

OBJECTIVE: You will be able to identify sources of plumbing jobs, explain the levels of the plumbing apprenticeship program and the program's educational requirements, and list qualifications for success in the plumbing trade. You will be able to describe characteristics of desirable employees, describe the process of getting and keeping a job, and explain the process of starting a small business.

Carefully study the chapter and then answer the following questions.

1. The five related areas in which a plumber might find employment are: Plumbing enterprises, Plumbing contractors associations, Plumbing union officials, governmental officials, and Plumbing _____.
   _____

2. _____ Plumbing _____ serve the needs of the community and offer the greatest opportunity as an entry point for most beginners.
   A. enterprises
   B. educators
   C. organizations
   D. contractors associations

3. _____ *True or False?* Plumbing enterprises include retailers as well as plumbing installation and repair services.

4. _____ Which of the following listed are functions of government agencies that hire plumbers?
   A. License plumbers.
   B. Review plans.
   C. Issue permits.
   D. All the above.

5. _____ *True or False?* Plumbing educators generally are master plumbers with many years of experience.

6. _____ Plumbing _____ provide benefits to the plumber, and are the source of employees for numerous contractors.
   A. unions
   B. associations
   C. certifications
   D. licenses

7. _____ An apprentice plumber goes through an educational program that is generally a minimum of _____ years long.
   A. three
   B. four
   C. five
   D. six

8. _____ *True or False?* Journeymen plumbers are qualified to be plumbing contractors.

9. _____ Master plumbers are generally required to work as journeymen for _____ years before taking the exam for a master's license.
   A. two
   B. three
   C. five
   D. ten

Copyright Goodheart-Willcox Co., Inc.
May not be reproduced or posted to a publicly accessible website.

10. _____ _____ are responsible for directing the work of a small group of workers.
    A. Supervisors
    B. Superintendents
    C. Master plumbers
    D. Contractors

11. _____ _____ oversee large plumbing jobs and generally have several supervisors working under their direction.
    A. Supervisors
    B. Superintendents
    C. Master plumbers
    D. Estimators

12. _____ _____ are responsible for estimating the material and actual cost of a job.
    A. Supervisors
    B. Superintendents
    C. Master plumbers
    D. Estimators

13. _____ Plumbing _____ are master plumbers who generally hire apprentices, journeymen, and other master plumbers to work for them.
    A. estimators
    B. supervisors
    C. contractors
    D. supply dealers

14. _____ The most important function of a plumbing supply dealer is to _____ needed materials to the jobsite at the time they are needed.
    A. order
    B. provide estimates for
    C. deliver
    D. suggest

15. _____ *True or False?* One of the most important skills that a plumber must acquire is the ability to visualize the completed plumbing system before beginning work on it.

16. _____ *True or False?* Only journeymen plumbers are eligible to obtain plumbing permits.

17. _____ The three-part foundation of basic skills, thinking skills, and _____ qualities can be considered fundamental to everyone's education.
    A. personal
    B. healthy
    C. physical
    D. speaking

18. _____ The five skills essential for success in the workplace are reading, writing, _____, listening, and speaking.
    A. critical thinking
    B. mathematics
    C. science
    D. punctuality

19. _____ The mathematics-related skill that is most frequently used by plumbers is _____.
    A. multiplication/division
    B. geometry
    C. measurement
    D. algebra

20. _____ Receiving, attending to, interpreting, and responding to verbal messages are all a part of _____ skills.
    A. speaking
    B. writing
    C. listening
    D. caring

Name _____

21. _____ Avoid foul, sexist, prejudicial, and _____ language in the workplace.
    A. opinionated
    B. inflammatory
    C. loud
    D. negative

22. _____ The six thinking skills identified by the Labor Department's report are reasoning, decision making, _____, problem solving, creative thinking, and knowing how to learn.
    A. critical thinking
    B. reading
    C. speaking
    D. visualization

23. _____ Which of the following is *not* a personal quality identified by the Labor Department?
    A. Integrity
    B. Sociability
    C. Efficiency
    D. Responsibility

24. Three things a plumber should do if he or she makes a mistake on the job are to _____ his or her mistake to a supervisor, correct the mistake, and try not to make the same mistake again.
    _____

25. A plumber's confidence in his or her ability enables the plumber to proceed with the work in an efficient and _____ manner.
    _____

26. _____ A plumber who has the ability to get along with diverse groups of people exhibits the personal quality of _____.
    A. integrity
    B. responsibility
    C. sociability
    D. self-management

27. _____ *True or False?* Self-management involves exercising self-control or self-discipline.

28. _____ Integrity involves being _____.
    A. honest
    B. trustworthy
    C. dependable
    D. All of the above.

29. _____ Five resources that a plumber must effectively manage are time, _____, tools, materials, and supplies.
    A. equipment
    B. apprentices
    C. clients
    D. teamwork

30. _____ *True or False?* Effectively using tools and equipment begins with selecting the appropriate tool for the task.

Match each of the following words with the statement that describes it in questions 31–38.

31. ____ Punctual
32. ____ Dependable
33. ____ Honesty
34. ____ Motivation
35. ____ Technical ability
36. ____ Leadership
37. ____ Quality of work
38. ____ Cooperative

A. Gives a fair day's work
B. Works willingly with other plumbers
C. Arrives on time and ready to work
D. Follows through on work assignments
E. Work consistently meets employer's standard
F. Truthful with the supervisor and coworkers
G. Guides others toward the completion of assigned work
H. Performs the work with limited supervision

39. ____ *True or False?* A family member is an example of a good reference.

40. ____ Finding employment opportunities through other people, such as friends, is called ____.
    A. referencing
    B. networking
    C. job hunting
    D. connecting

41. ____ *True or False?* Plumbing suppliers are a source of job leads.

42. ____ *True or False?* A plumber's résumé provides documentation of work experience and may contain plans, sketches, photographs, and a list of jobs the plumber has completed.

43. ____ Before starting a business, working as a part-time self-employed plumber allows you to obtain management experience, develop a client base, establish a reputation, arrange accounts, report ____, and do general record keeping.
    A. employee documents
    B. work completed
    C. expenses
    D. All of the above.

44. ____ A business plan typically includes a description of the products and services to be marketed, a marketing plan, resources required to operate the business, and a ____.
    A. list of employees
    B. financial plan
    C. list of personal expenses
    D. sales territory

Name _____ Date _____ Class _____

CHAPTER 4: Plumbing History

OBJECTIVE: You will be able to describe examples of plumbing developments from ancient times; identify contributions made by the Romans; trace the development of plumbing in Western Europe and the United States; and describe the development of plumbing fixtures, piping materials, and tools. You will be able to describe the contribution of laws, codes, and organizations to the development of plumbing.

Carefully study the chapter and then answer the following questions.

1. The history of plumbing can be traced back __5,000__ years to the Babylonians.
   _____

2. __B__ The ancient Romans were famous for their _____ and public baths.
   A. piping
   B. aqueducts
   C. hygiene
   D. architecture

3. _____ *True or False?* In Western Europe, sewers were constructed in some areas to help control storm water. (True circled)

4. _____ *True or False?* London's cholera epidemic in the early 1800s was traced to the lack of personal hygiene. (False circled)

5. _____ The _____ Empire was the first to provide large numbers of its citizens with plumbing.
   A. Greek
   B. Roman
   C. Babylonian
   D. British

6. _____ *True or False?* The Black Plague was directly connected to the lack of sanitary waste disposal. (True circled)

Match the following terms and descriptive phrases for questions 7–9.

7. __C__ Disease traced to contaminated drinking water   A. Dark Ages
8. __A__ 400–1000 CE   B. Black Plague
9. __B__ Caused the death of 25 million people   C. Cholera

10. __D__ A _____ epidemic in 1854 killed five percent of Chicago's population.
    A. sewage
    B. dysentery
    C. typhoid
    D. cholera

11. _____ *True or False?* In the mid-1800s, Chicago's water supply became contaminated from garbage that was being dumped into the Chicago River. (False circled)

12. _____ A New York state law passed in 1881 required that plumbers be _____.
    A. trained
    B. registered
    C. union members
    D. None of the above.

13. __A__ Early attempts to install plumbing indoors failed because sewer gases escaped into the house even though _____ were installed.
    A. ventilation systems
    B. insulated walls
    C. pipes
    D. traps

14. __C__ The general idea of _____ drain/waste piping systems was first used in New York in 1874.
    A. building
    B. cleaning
    C. venting
    D. None of the above.

15. __A__ What size pipe was first used to vent traps in New York City?
    A. 1/2″
    B. 1″
    C. 1-1/2″
    D. 2″

16. _____ The Greeks' major contribution to plumbing was the development of bathtubs with _____.
    A. faucets
    B. cold water
    C. drains
    D. hot water

17. __B__ Bitumen is a natural type of _____ that was used to make clay floors waterproof in about 3000 BCE.
    A. clay
    B. asphalt
    C. metal
    D. polymer

18. _____ *True* or *False?* The Romans used lead to make pipe and line sinks and other fixtures.

19. __C__ Plumbum is the Latin word for _____.
    A. plumber
    B. pipe
    C. lead
    D. bathhouse

20. __A__ In 1562, _____ pipe was used to supply water to a fountain in Germany.
    A. cast-iron
    B. lead
    C. clay
    D. stone

21. __B__ Early bathtubs were made of _____ in the shape of a shoe.
    A. iron
    B. copper
    C. lead
    D. tin

22. __B__ Evidence of _____ has been found in the ruins of Egyptian cities that date back to 1350 BCE.
    A. bathtubs
    B. showers
    C. sinks
    D. water closets

Name _____

23. _____ In 1775, _____ of England patented the water closet that became the prototype for today's toilets.
    A. Alexander Cummings
    B. Alexander Bell
    C. Sir John Harington
    D. George Jennings

24. _____ In 1906, _____ applied for a patent for the flush valve, which made possible the tankless toilet used in most public restrooms.
    A. John Harington
    B. Joseph Bramah
    C. William E. Sloan
    D. Thomas Twyford

25. _____ Boston set up the first waterworks in the United States in 1652 to supply water to wharves and buildings for _____ and domestic use.
    A. drainage
    B. firefighting
    C. storing
    D. None of the above.

26. _____ The automatic storage water heater was invented _____.
    A. in the 1930s
    B. by an employee of the Kohler Co.
    C. by Edwin Ruud
    D. All of the above.

27. _____ *True or False?* New York City became the first city in the United States to select cast iron for water mains.

28. __B__ Galvanized pipe is coated with _____ metal to prevent rust.
    A. aluminum
    B. zinc
    C. copper
    D. molten iron

29. _____ *True or False?* PVC pipe began to be practically manufactured during the 19th century.

30. The first compact _____ was called the Ditch Witch™.
    _____

31. In which year were each of the following federal laws enacted?

    A. _____ Occupational Safety and Health Act

    B. _____ Safe Drinking Water Act

    C. _____ Americans with Disabilities Act

32. _____ Water efficiency regulations for plumbing fixtures were established by the _____.
    A. Hoover Code
    B. International Residential Code for One- and Two-Family Dwellings
    C. Uniform Plumbing Code
    D. National Plumbing Standard

33. When did plumbing as we know it in the United States begin to develop?
    1920's

Notes

Name _____ Date _____ Class _____

CHAPTER 5
Leveling Instruments

OBJECTIVE: You will be able to explain the operation of the builder's level and laser level and describe how they are used to find levels and properly slope drainage pipe.

Carefully study the chapter and then answer the following questions.

1. _____ Plumbers may use a _____ for leveling.
   A. level
   B. straightedge
   C. chalk line
   D. All the above.

2. _____ A builder's level is also known as a _____.
   A. surveyor's level
   B. transit
   C. line level
   D. Both A and B.

3. _____ A builder's level mounts on a _____.
   A. bipod
   B. tripod
   C. bench
   D. pipe vise

4. _____ A builder's level rotates _____ degrees.
   A. 45
   B. 90
   C. 180
   D. 360

5. _____ A _____ is a graduated stick used with a builder's level to find elevations.
   A. pole rod
   B. stadia rod
   C. tape
   D. framing square

6. _____ Using a builder's level, a plumber has a first reading of 6'-0", a second reading of 9'-4", and a length of run of 100'. Compute the rate of fall.
   A. 1/3" fall/foot of run
   B. 3 1/3" fall/foot of run
   C. 33" fall/foot of run
   D. 33 1/3" fall/foot of run

7. _____ *True or False?* The horizontal crosshair in the telescope sight indicates the point at which the reading should be made.

8. _____ A(n) _____ level is self-adjusting to maintain a horizontal line of sight.
   A. pipe laser
   B. laser
   C. builder's
   D. automatic

Copyright Goodheart-Willcox Co., Inc.
May not be reproduced or posted to a publicly accessible website.

Match the parts of the builder's level with the proper part name for questions 9–18.

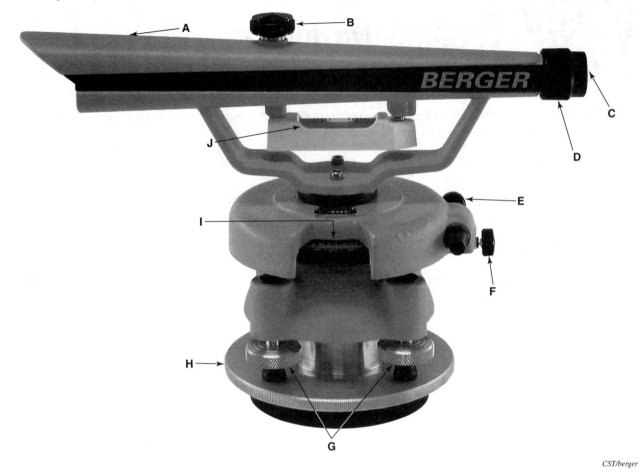

CST/berger

9. _____ Telescope
10. _____ Leveling screws
11. _____ Eyepiece
12. _____ Telescope focus knob
13. _____ Eyepiece focus

14. _____ Horizontal motion screw
15. _____ Telescope level
16. _____ Horizontal motion tangent screw
17. _____ Horizontal circle Vernier
18. _____ Base plate

19. A(n) _____ is an instrument that amplifies, or strengthens, light, projecting it as a thin beam.

20. _____ *True or False?* Long-term exposure to a laser light beam may cause eye injury.

21. _____ *True or False?* A worker does *not* need to be qualified to operate laser equipment.

22. _____ *True or False?* When a laser is being operated, beam shutters and caps should be used.

23. _____ When a laser is left unattended, it should be _____.
    A. turned off
    B. recharged
    C. left on
    D. None of the above.

24. _____ During installation of a sewer pipe using a laser, the laser beam is projected through the pipe until it strikes a(n) _____.
    A. block
    B. fitting
    C. end cap
    D. beam target

Name _____ Date _____ Class _____

CHAPTER 6
Mathematics for Plumbers

OBJECTIVE: You will be able to read a rule, add and subtract fractions and whole numbers, find area and volume, and convert measurements.

Carefully study the chapter and then answer the following questions.

1. _____ The _____ and steel tape are the basic measuring tools used by plumbers.
    A. folding rule
    B. plumber's rule
    C. meter stick
    D. None of the above.

2. Identify the measurements indicated on the following rule.

    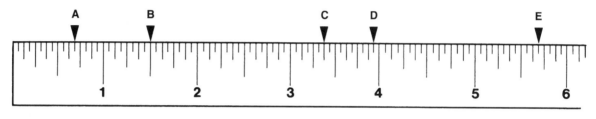

    Rule

    Goodheart-Willcox Publisher

    A. _____
    B. _____
    C. _____
    D. _____
    E. _____

3. _____ To prevent costly errors when cutting materials, measurements should be _____.
    A. written down
    B. rechecked
    C. avoided
    D. rounded to the nearest whole number

4. _____ There are _____ 1/16s in 7/8.
    A. 6
    B. 12
    C. 14
    D. 16

5. _____ There are _____ 1/8s in 3/4.
    A. 4
    B. 6
    C. 8
    D. 10

Copyright Goodheart-Willcox Co., Inc.
May not be reproduced or posted to a publicly accessible website.

6. Solve the following problems.

   A. 14 1/2 + 2 1/2 = _____

   B. 22 3/4 − 3 1/4 = _____

   C. 8 5/8 + 2 3/16 = _____

   D. 10 3/4 + 1 3/8 = _____

   E. 12 1/2 − 2 3/4 = _____

7. A measurement of 12′ converts to _____″.

_____

8. A measurement of 197′ converts to _____″.

_____

9. _____ In the fraction 1/2, the number 2 is the _____.
   A. numerator
   B. denominator
   C. dimension
   D. mixed number

For questions 10–13, use the drawing to answer the question. Round to the nearest hundredth.

10. In pipe layout I, the distance of AB is _____″.

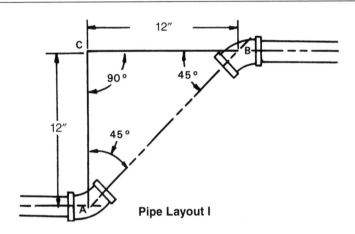

Pipe Layout I

Goodheart-Willcox Publisher

11. In pipe layout II, the distance of AB is _____″.

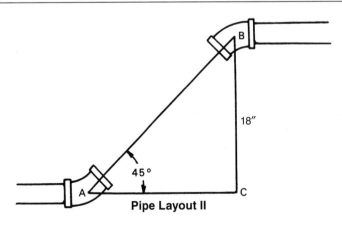

Pipe Layout II

Goodheart-Willcox Publisher

Name _____

12. In pipe layout III, the distance of AC is _____″.

_____

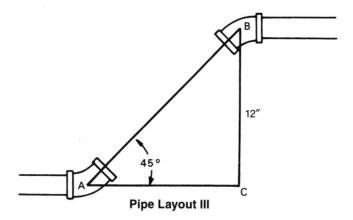

Pipe Layout III

13. In pipe layout IV, identify the run, offset, and travel.

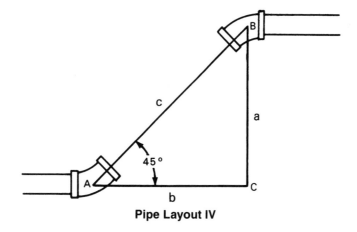

Pipe Layout IV

    A. Run _____

    B. Offset _____

    C. Travel _____

14. _____ The surface area of a square or a rectangular surface can be computed by multiplying the length by the _____.
    A. height
    B. diameter
    C. width
    D. volume

For questions 15–17, use the drawing to answer the question. Round to the nearest hundredth.

15. The area of the following rectangle is _____ square feet.

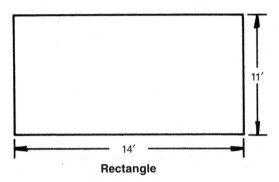

Rectangle

Goodheart-Willcox Publisher

16. The area of the following circle is _____ square inches.

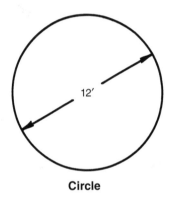

Circle

Goodheart-Willcox Publisher

17. The volume of the following rectangular tank is _____ cubic feet.

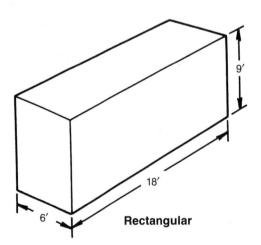

Rectangular

Goodheart-Willcox Publisher

Name _____ Date _____ Class _____

Use the figure below to answer questions 18–19. Round to the nearest hundredth.

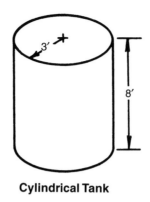

Cylindrical Tank

Goodheart-Willcox Publisher

18. The volume of the cylindrical tank is _____ cubic feet.

_____

19. If 7.48 gallons of water equals one cubic foot, then _____ gallons of water can be held in the cylindrical tank.

_____

20. _____ To convert cubic inches to gallons, divide by _____.
    A. 7.48
    B. 25.4
    C. 231
    D. 1,728

21. Six gallons equals _____ liters.

_____

22. There are _____ meters in 30 feet.

_____

23. _____ To change from one metric unit to another, multiply or divide by multiples of _____.
    A. 2
    B. 10
    C. 12
    D. 100

24. _____ The metric unit used for long distances is the _____.
    A. meter
    B. centimeter
    C. foot
    D. mile

25. _____ The metric unit used for very small measurements is the _____.
    A. centimeter
    B. inch
    C. decimeter
    D. millimeter

26. _____ The metric unit used for dry volume is the cubic _____.
    A. centimeter
    B. meter
    C. inch
    D. foot

27. Convert the following measurements. Round to the nearest hundredth.
    A. 16″ = _____ millimeter(s)
    B. 12″ = _____ meter(s)
    C. 9″ = _____ centimeter(s)
    D. 478 cubic inches = _____ liter(s)
    E. 3.75 cubic feet = _____ liter(s)
    F. 24 cubic inches = _____ cubic feet
    G. 20 cubic feet = _____ cubic yard(s)
    H. 12 cubic inches = _____ gallon(s)
    I. 5 cubic feet = _____ gallon(s)
    J. 35 cubic feet = _____ cubic meter(s)

28. The _____ is the metric unit for force.
    A. pascal
    B. degree
    C. newton
    D. psi

29. The _____ scale is used to measure temperatures in the super-cold range of (–273°C and below).
    A. Celsius
    B. Fahrenheit
    C. US customary
    D. Kelvin

30. Convert the following temperatures. Round to the nearest hundredth.
    A. 20°C = _____ °K
    B. 220°F = _____ °C
    C. 25°C = _____ °F
    D. 298°K = _____ °C
    E. 72°F = _____ °C

Name _____ Date _____ Class _____

CHAPTER 7
Hydraulics and Pneumatics

OBJECTIVE: You will be able to explain and apply characteristics of fluids under pressure, apply practical methods of computing useful pressures, and demonstrate undesirable characteristics of gases in incorrectly designed drainage systems.

Carefully study the chapter and then answer the following questions.

1. __T__ *True or False?* Hydraulics is the study of the characteristics of fluids.
2. __B__ A cubic foot of water weighs _____ pounds.
   A. 14.8
   B. 62.4
   C. 82.4
   D. 100
3. __F__ *True or False?* Water pressure decreases as the depth of water increases.
4. __B__ If water is stored in a tower 175 feet tall, the water pressure at the base of the tower is _____ psi.
   A. 2.8
   B. 910
   C. 10,850
   D. 10,920
5. _____ _____ is the pressure available at some point in the water system, and is measured by the depth of a column of water above the point where the measurement is taken.
   A. Hydraulics
   B. Pressure head
   C. Friction loss
   D. Pressure per square inch
6. _____ According to scientific studies, a column of water 1′ high produces a pressure of _____ psi.
   A. 0.43
   B. 0.86
   C. 1.20
   D. 1.40
7. _____ *True or False?* Plumbers must consider the effects of friction on water pressure and flow rates.
8. _____ *True or False?* Streamlined flow produces considerable friction, while turbulent flow produces little friction.
9. _____ Compared to schedule 40 plastic pipe, grade L copper pipe has _____ head loss.
   A. more
   B. less
   C. equal
   D. None of the above.
10. _____ *True or False?* Water hammer can cause pipes to vibrate and possibly burst.
11. _____ *True or False?* Pneumatics is the study of compressible gases.
12. __C__ A(n) _____ bend in a low-pressure piping system is likely to cause an air lock that blocks the flow through the pipe.
    A. 90°
    B. downward
    C. upward
    D. 45°

13. ____ *True or False?* Under normal conditions, both the sanitary and storm sewer are completely filled with water.

14. ____ *True or False?* In the following illustration, the DWV piping system shown for Building B is better than the system in Building A.

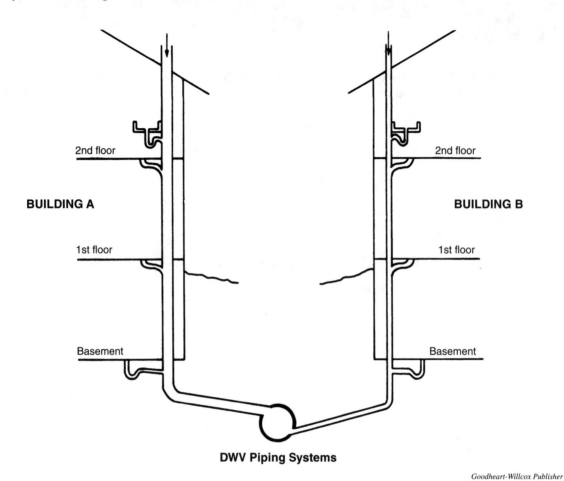

DWV Piping Systems

Goodheart-Willcox Publisher

15. __C__ Commercial airplanes and cruise ships generally use a ____ to pull waste through the waste piping.
    A. DWV piping system
    B. shallow well pump
    C. vacuum system
    D. trapped air system

Name _____ Date _____ Class _____

CHAPTER 8
Print Reading and Sketching

OBJECTIVE: You will be able to identify the plumbing symbols and abbreviations, interpret various kinds of plans, and take dimensions of drawings in inches and feet. You will be able to scale drawings and prepare two- and three-dimensional piping sketches.

Carefully study the chapter and then answer the following questions.

1. _____ *True or False?* Scale drawings may use fractions of inches to represent feet.

2. _____ Drawings are frequently referred to as _____.
   A. prints
   B. blueprints
   C. master plans
   D. Both A and B.

3. _____ A _____ may use prints.
   A. building official
   B. contractor
   C. supplier
   D. All of the above.

Match the plan description with the proper plan for questions 4–10.

4. _____ Plot plan
5. _____ Elevation drawings
6. _____ Floor plan
7. _____ Foundation plan
8. _____ Detail drawings
9. _____ Electrical plan
10. _____ Heating plan

A. Construction of walls, stairs, and doors
B. Describes arrangement of rooms
C. Describes size and shape of footing
D. Location of structure on the lot
E. Location of service entry and distribution panel
F. Exterior appearance
G. Describes placement of heating system components

11. _____ _____ are a set of instructions that provide information about materials and work quality.
    A. Blueprints
    B. Specifications
    C. Catalogs
    D. None of the above.

12. _____ *True or False?* A section view indicates the horizontal distances critical to the installation of DWV piping.

13. _____ Dimensions on architectural drawings are usually given in _____.
    A. meters
    B. feet and inches
    C. inches only
    D. feet only

14. _____ A distance of 83 inches is commonly written as _____.
    A. 83′
    B. 6′-11″
    C. 5′-11″
    D. None of the above.

15. _____ _____ are symbols used to indicate the limits of a particular dimension.
    A. Arrowheads
    B. Dots
    C. Diagonal lines
    D. All of the above.

16. Find, in the simplest form, the total dimension of 6'-10", 3'-3", and 8'-11".

    A. Total feet _____

    B. Total inches _____

    C. Simplified _____

Match the symbols with plumbing fixtures, appliances, and mechanical equipment for questions 17–26.

17. _____ Dry well
18. _____ Range
19. _____ Sump pit
20. _____ Water heater
21. _____ Vacuum outlet
22. _____ Hose bibb
23. _____ Floor drain
24. _____ Dishwasher
25. _____ Built-in cooking top
26. _____ Water softener

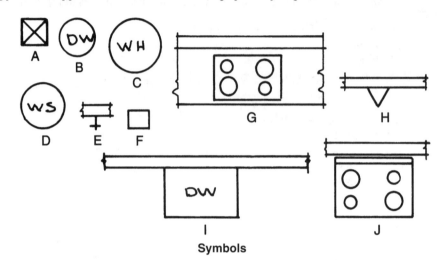

Symbols

Goodheart-Willcox Publisher

Identify the abbreviations for questions 27–39.

A. FD   E. HB   H. LT   K. LAV
B. HW   F. GI   I. PLAS   L. CW
C. CI   G. PLBG   J. WC   M. DW
D. CO

27. _____ Cast iron
28. _____ Clean out
29. _____ Cold water
30. _____ Hot water
31. _____ Dishwasher
32. _____ Floor drain
33. _____ Hose bibb
34. _____ Galvanized iron
35. _____ Plastic
36. _____ Lavatory
37. _____ Plumbing
38. _____ Laundry tray
39. _____ Water closet

30 Modern Plumbing Lab Workbook

Name _____

40. _____ The following symbol represents a _____ connection.

    A. soldered or cemented
    B. bell and spigot
    C. screwed
    D. None of the above.

41. _____ The following symbol represents a(n) _____.

    A. Reducing elbow
    B. Elbow—90°
    C. Elbow—45°
    D. None of the above.

42. _____ *True or False?* To find the correct scale of a drawing, the title block is referenced.

43. _____ Building plans are customarily drawn to a _____ scale.
    A. 1/2 inch = 1 foot
    B. 1/4 inch = 1 foot
    C. 1 mm = 50 mm
    D. Both B and C.

44. _____ *True or False?* A recessed tub has a wall on two sides.

45. Find the dimension of each opening shown in the following illustrations.

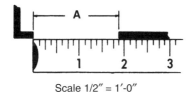 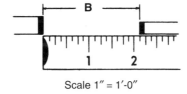 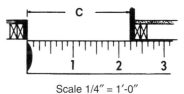

    A. _____
    B. _____
    C. _____

46. _____ The three types of piping sketches are the _____, the two-dimensional, and the isometric.
    A. plan view
    B. architectural
    C. metric
    D. scale

47. _____ *True or False?* An isometric sketch is the simplest type of piping sketch.

48. _____ The _____ is a piping sketch that is frequently drawn at the jobsite.
    A. plan view
    B. two-dimensional riser diagram
    C. isometric
    D. All of the above.

49. Using 1, 2, 3, and 4, indicate the order of the steps in making a plan view sketch.
    A. _____ Darken lines and add notes.
    B. _____ Lightly draw pipe.
    C. _____ Locate known points.
    D. _____ Indicate fitting symbols.

50. _____ In an isometric axis, converging lines form angles of _____° or _____° from horizontal.
    A. 90; 30
    B. 90; 45
    C. 120; 30
    D. 120; 45

51. _____ *True or False?* Isometric sketches are helpful in illustrating more complex piping systems.

52. _____ _____ grid paper is extremely helpful in making an isometric sketch.
    A. Horizontal
    B. Vertical
    C. Isometric
    D. None of the above.

Name _____ Date _____ Class _____

CHAPTER 9
Green Construction

OBJECTIVE: You will be able to describe the purpose of green construction and identify green plumbing devices that can reduce water consumption.

Carefully study the chapter and then answer the following questions.

1. A way of building to reduce the impact of construction on the environment is called _____ construction.
   _____

2. _____ *True or False?* Using processes that are environmentally responsible and resource-efficient is a principal means of achieving green construction.

3. _____ To be truly *green*, materials should be _____.
   A. green in color
   B. practical
   C. sustainable
   D. low in cost

4. _____ The easiest way for plumbers to reduce greenhouse gases is to purchase _____ materials for construction projects.
   A. cost-effective
   B. locally available
   C. limited
   D. recycled

5. _____ _____ are powerful machines that force heated copper through round dies to produce pipe.
   A. Billets
   B. Extruders
   C. Trenchers
   D. None of the above.

6. _____ *True or False?* One of the keys to minimizing the amount of pollution created by excavation work is optimizing worker schedules.

7. _____ To conserve water, always select toilets that use _____ or less per flush.
   A. 1.6 gallons
   B. 3.5 gallons
   C. 4 gallons
   D. None of the above.

8. _____ *True or False?* Repairing leaking faucets will reduce wasteful use of water.

9. _____ A steady drip can waste as much as _____ gallons per year.
   A. 1,000
   B. 2,000
   C. 8,000
   D. 10,000

10. Each person living in a home uses an average of _____ gallons of water per day.
    _____

Copyright Goodheart-Willcox Co., Inc.
May not be reproduced or posted to a publicly accessible website.

11. _____ _____ water systems use outflow from lavatories, showers, and tubs to flush toilets or for irrigation.
    A. Gray
    B. Recycled
    C. LEED-approved
    D. None of the above.

12. _____ _____ toilets use less water to flush liquid waste than is needed for solid waste.
    A. Single-flush
    B. Dual-flush
    C. Composting
    D. All of the above.

13. _____ The International Code Council (ICC) 700 _____ is a nationally recognized program that rates green buildings.
    A. LEED certification
    B. National Green Building Standard
    C. ENERGY STAR®
    D. Environmental Protection Agency (EPA)

14. _____ The ICC 700 National Green Building Standard rating system rates _____ buildings only.
    A. commercial
    B. residential
    C. energy-efficient
    D. water-efficient

15. LEED rates _____ building types.
    _____

16. _____ *True or False?* Rainwater that has been collected and stored can be used to water plants.

17. _____ _____% of the Earth's surface is covered by water.
    A. 25
    B. 70
    C. 85
    D. 99

18. _____ Less than _____% of accessible water is available for human consumption.
    A. 0.1
    B. 0.5
    C. 0.75
    D. 1

19. _____ *True or False?* Water is *not* a scarce resource.

20. What is a life-cycle assessment?
    _____
    _____
    _____

Name _____ Date _____ Class _____

CHAPTER 10
Building and Plumbing Codes

OBJECTIVE: You will be able to explain zoning laws and building codes and how they are administered and enforced. You will be able to list points a plumbing code should cover and be able to apply code requirements to a plumbing installation.

Carefully study the chapter and then answer the following questions.

1. _____ *True or False?* Minimum standards in buildings are specified to provide for the health and safety of the people.

2. _____ _____ are laws that regulate the type of structure that can be built in a given area.
   A. Zoning laws
   B. Plumbing codes
   C. Electrical codes
   D. Building standards

3. _____ *True or False?* Within zoning laws, residential, office, and industrial activities are *not* separated.

4. _____ Building codes control the _____.
   A. building materials
   B. quality of work
   C. electrical service
   D. All of the above.

5. _____ *True or False?* Plumbing and electrical codes were originally developed independently of building codes.

6. _____ *True or False?* The enforcement of the plumbing code is generally delegated to the National Plumbing Commission.

7. _____ When applying for a building permit, the contractor must submit two copies of _____ and _____ to the building inspector.
   A. plans; regulations
   B. codes; blueprints
   C. plans; specifications
   D. sepias; codes

8. _____ *True or False?* After a permit is obtained, a construction project will be inspected only after its completion.

9. _____ *True or False?* If work being done does not meet the codes, building officials can obtain a court order to stop construction.

10. _____ To improve the quality of codes and provide a degree of standardization, _____ codes were developed.
    A. plumbing
    B. model
    C. federal
    D. local

11. _____ _____ are involved in the development of plumbing codes.
    A. Local governments
    B. Architects
    C. Plumbers
    D. All of the above.

12. _____ *True or False?* Before a model plumbing code can be enforced, it must be adopted by the local government.

Match the model code with its sponsoring organization for questions 13–16.

13. _____ International Residential Code for One- and Two-Family Dwellings

14. _____ BOCA National Building Code

15. _____ Standard Building Code

16. _____ Uniform Building Code

A. Building Officials and Code Administrators International
B. International Code Council
C. Southern Building Code Conference International, Inc.
D. International Conference of Building Officials

17. _____ *True or False?* Plumbers have the authority to modify the plumbing code to meet their needs.

18. _____ The two ICC codes that are most important for plumbers are the International Residential One- and Two-Family Dwelling Code and the _____ Plumbing Code.
    A. National
    B. International
    C. Commercial
    D. Model

Name _____ Date _____ Class _____

CHAPTER 11
Soldering, Brazing, and Welding

OBJECTIVE: You will be able to identify solders, brazing filler metals, and fluxes used for soldering and brazing copper pipe and fittings. You will be able to describe the processes by which pipe and fittings are joined.

Carefully study the chapter and then answer the following questions.

1. _____ _____ is the method of using heat to form joints between two metallic surfaces using a nonferrous filler metal.
   A. Flaring
   B. Swagging
   C. Soldering
   D. Lasering

2. _____ A nonferrous metal does *not* contain _____.
   A. aluminum
   B. iron
   C. copper
   D. flux

3. _____ *True or False?* Capillary attraction is the amount of attraction a metal has to a magnet.

4. _____ *True or False?* Solder composed of 50% tin and 50% lead should be used for potable water supply piping.

5. _____ Which of the following materials is *not* typically brazed with hard solders?
   A. Cast iron.
   B. Brass.
   C. Aluminum.
   D. Steel.

6. _____ _____ is the process of picking up oxygen that produces tarnish and rust in metals.
   A. Soldering
   B. Brazing
   C. Oxidation
   D. Welding

7. _____ _____ fluxes are best suited for plumbing work.
   A. Noncorrosive
   A. Corrosive
   B. Less corrosive
   C. None of the above.

8. _____ *True or False?* It is important to clean copper pipe and fittings before applying flux.

9. _____ When lighting a torch, _____.
   A. use a spark lighter
   B. direct tip away from you
   C. direct tip away from flammable materials
   D. All of the above.

10. _____ *True or False?* The solder should be melted in the flame of the torch.

11. _____ Brazing is done at temperatures above _____°F.
    A. 200
    B. 400
    C. 800
    D. 1000

Copyright Goodheart-Willcox Co., Inc.
May not be reproduced or posted to a publicly accessible website.

Match the cleaning tools with their proper name.

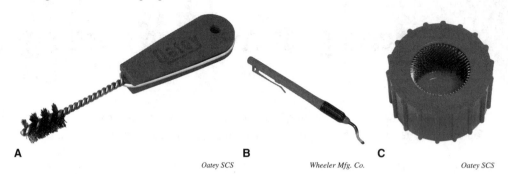

A   B   C
   Oatey SCS   *Wheeler Mfg. Co.*   *Oatey SCS*

12. _____ Tube brush

13. _____ Fitting brush

14. _____ Reamer

15. _____ *True or False?* Brazing is the same process as braze welding.

16. _____ *True or False?* The strength of a joint comes from the ability of the silver braze to flow into the porous grain structure of the base metal.

17. _____ To ensure an excellent bond when brazing, the clearance gap between pipe and fitting must be only _____″ to _____″.
    A. .002, .004
    B. .002, .005
    C. .003, .004
    D. .003, .006

Match each of the braze materials with its proper description for questions 18–22.

18. _____ Nickel

19. _____ Copper-phosphorus

20. _____ Silver

21. _____ Copper and copper-zinc

22. _____ Aluminum-silicon

A. Used for joining copper and copper alloys
B. Used for brazing aluminum
C. Used for extreme heat and corrosion resistance
D. Used for brazing all ferrous and nonferrous metals, except aluminum
E. Used for brazing iron and steel

23. _____ To braze 1-1/2″ or 2″ pipe, a No. 7 torch tip is recommended, which requires _____ psi oxygen pressure and 7 psi acetylene pressure.
    A. 5
    B. 7
    C. 8
    D. 10

24. _____ *True or False?* A neutral flame should *never* be used for brazing.

25. _____ A carburizing flame is the result of an excess of _____.
    A. oxygen
    B. acetylene
    C. carbon
    D. Both A and B.

26. _____ Welding in the plumbing industry is generally limited to repair work on _____ pipe systems.
    A. cast-iron
    B. thermoplastic
    C. residential
    D. commercial

Name _____ Date _____ Class _____

CHAPTER 12
Excavating

OBJECTIVE: Explain the importance of locating and protecting existing underground utilities. Explain the importance of protecting workers while trenching. Develop a list of general safety rules regarding the use of excavating tools and machines.

Carefully study the chapter and then answer the following questions.

1. _____ *True or False?* All underground utilities must be located and protected before digging begins.

2. _____ Excavating work associated with plumbing is done with _____.
   A. backhoes
   B. front loaders
   C. trenchers
   D. All of the above.

3. _____ *True or False?* Trenchers or small backhoes can be used to produce narrow trenches.

4. _____ _____ requires that excavations and trenches over _____′ deep be guarded with shoring.
   A. NFPA; 5
   B. ASTM; 5
   C. OSHA; 10
   D. OSHA; 5

5. _____ _____ may be used in place of shoring or sloping.
   A. Sleeves
   B. Trench boxes
   C. Steel pipe
   D. Walers

6. _____ *True or False?* Vertical strut shoring is generally used in very unstable soils.

7. _____ _____ shovels are used for digging a narrow trench and loosening compacted earth.
   A. Round-point
   B. Trench
   C. Square-point
   D. Drain spade

8. _____ If the ground is too hard to be loosened with a shovel, you may need to use a spud bar, pick, or _____.
   A. spade
   B. mattock
   C. trencher
   D. backhoe

9. _____ *True or False?* When you complete the swing of a pick/mattock, the point of the tool will be parallel to the surface being struck.

10. _____ *True or False?* When striking dense material with a pick/mattock, you and others around you must wear eye protection.

11. _____ Removing large quantities of dense material is best accomplished with a chipping _____.
    A. compressor
    B. chisel
    C. hammer
    D. pick

12. _____ Pneumatic breakers or chipping hammers are commonly called _____.
    A. jackhammers
    B. rotary hammers
    C. ball peen hammers
    D. sledgehammers

13. _____ A _____ is used at the work site to provide a high volume of compressed air for pneumatic tools.
    A. portable motor
    B. portable compressor
    C. portable hammer
    D. chipping hammer

14. _____ _____ are used to install pipe without cutting through sidewalks, driveways, and streets.
    A. Jackhammers
    B. Small backhoes
    C. Directional drills
    D. Hammer drills

15. _____ _____ can be placed in the bottom of a trench to support the pipe at the required slope.
    A. Gravel
    B. Cement
    C. Sand
    D. None of the above.

16. What is the cut-and-cover technique?

_____

_____

_____

Name _____ Date _____ Class _____

CHAPTER 13
Water Supply Systems

OBJECTIVE: You will be able to list the types of wells, describe the various types of water pumps, and explain the operation of a pressure tank.

Carefully study the chapter and then answer the following questions.

1. _____ The _____ cycle makes it possible for people to repeatedly use water.
   A. rain
   B. water
   C. evaporation
   D. None of the above.

2. _____ *True or False?* All groundwater is safe for human consumption.

3. _____ The _____ is responsible for monitoring water sources so they meet standards.
   A. EPA
   B. Safe Standards Institute
   C. Potable Water Agency
   D. None of the above.

4. _____ *True or False?* Sulfur may be added to groundwater to purify it.

5. _____ A(n) _____ is a potential source of well contamination.
   A. underground fuel tank
   B. livestock feedlot
   C. waste disposal site
   D. All of the above.

6. _____ Which of the following is *not* one of the four most common wells?
   A. Dug.
   B. Driven.
   C. Compressed.
   D. Bored.

7. _____ *True or False?* Dug wells are seldom more than 20' deep.

8. _____ Driven wells are made by forcing a(n) _____ into the ground.
   A. percussion rig
   B. rotary rig
   C. well point
   D. earth auger

9. _____ The two most commonly used methods of drilling wells are the _____ and the rotary drilling methods.
   A. percussion
   B. drilling
   C. excavating
   D. None of the above.

10. _____ *True or False?* The vertical distance that the water is to be lifted is an important factor in water pump selection.

11. _____ Water weighs _____ pounds per gallon.
    A. four
    B. six
    C. eight
    D. ten

Copyright Goodheart-Willcox Co., Inc.
May not be reproduced or posted to a publicly accessible website.

12. _____ *True or False?* When the peak demand varies greatly from the average demand, a small storage should be used.

13. _____ For a typical plumbing fixture, a minimum water pressure of _____ pounds per square inch is required.
    A. 20
    B. 30
    C. 40
    D. 50

14. _____ The total distance the water must be pumped, the number of fittings, and the speed at which the water is traveling through the pipe all contribute to _____.
    A. volume available
    B. friction loss
    C. static pressure
    D. available water source

15. _____ The atmosphere exerts a pressure of _____ psi on the surface of water.
    A. 11
    B. 12.77
    C. 14.7
    D. 30.7

16. _____ _____ pumps lift water by moving a piston back and forth in a cylinder.
    A. Centrifugal
    B. Jet
    C. Rotary
    D. Reciprocating

17. _____ To lift water, centrifugal pumps use a rapid spinning _____, which is a heavy disk mounted in the pump housing.
    A. ejector
    B. motor
    C. impeller
    D. rotor

18. _____ *True or False?* Jet pumps are difficult to use because of the number of moving parts in the well that are vulnerable to deterioration.

19. _____ *True or False?* Rotary pumps have a constant discharge rate and operate efficiently.

20. _____ Wells deeper than _____′ require a pumping device installed within a well casing.
    A. 6
    B. 12
    C. 15
    D. 25

21. _____ *True or False?* Each stage of a submersible centrifugal pump decreases water pressure.

22. _____ *True or False?* Local codes may require that pumps be installed by a licensed worker.

23. _____ *True or False?* A well must be sanitized to overcome any unsanitary condition caused by the drilling operation.

24. _____ To prevent galvanic corrosion, a _____ fitting should be installed between a submersible pump and the drop pipe if they are dissimilar metals.
    A. galvanized
    B. copper
    C. dielectric
    D. DWV

25. _____ Hydropneumatic tanks contain _____ under pressure.
    A. air
    B. water
    C. oil
    D. Both A and B.

Name _____

26. _____ Which of the following statements is true about hydropneumatic tanks?
    A. As water volume builds up in the tank, so does the pressure.
    B. Hydropneumatic tanks are economical for large water supply systems.
    C. Supercharge is the amount of water pressure in the tank.
    D. Both A and B.

27. _____ *True or False?* As a general rule for sizing pressure tanks, the usable capacity should be three times the pump capacity.

28. _____ *True or False?* The outside of the well casing and the drilled hole must be filled to prevent surface water from entering the well.

29. _____ To prevent freezing, a(n) _____ must be installed where water piping is kept below the frost line.
    A. well seal
    B. pitless adapter
    C. cement grout
    D. O-ring gasket

30. _____ *True or False?* Using a drill and tapping machine, it is possible to tap a water main without turning off the water.

Notes

Name _____ Date _____ Class _____

CHAPTER 14
Water Treatment

OBJECTIVE: You will be able to identify water contaminants, describe devices to treat water, and know which devices have an effect on each of the contaminants.

Carefully study the chapter and then answer the following questions.

1. _____ Bacteria that are considered harmful to humans are called _____.
   A. organic chemicals
   B. pathogens
   C. chlorine
   D. inorganic chemicals

2. _____ _____ of the water supply is the most widely used means of killing bacteria.
   A. Purification
   B. Calcification
   C. Chlorination
   D. None of the above.

3. _____ Heavy metals that may be found in water include _____.
   A. tetrachloride and toluene
   B. mercury and lead
   C. selenium and nitrates
   D. arsenic and magnesium

4. _____ A reddish stain on fixtures is produced by _____.
   A. turbidity
   B. high pH water
   C. iron oxide
   D. manganese

5. _____ *True or False?* Plumbers are called on to install and maintain water-treatment equipment.

6. _____ Water that is safe for human consumption is called _____ water.
   A. clean
   B. pure
   C. filtered
   D. potable

7. _____ *True or False?* Domestic animals, such as livestock, do *not* require potable water.

8. _____ *True or False?* The first step in the design of a treatment system is to conduct a test of the water.

9. _____ A feed pump may be used to introduce _____ into the water system.
   A. sodium hypochlorite
   B. ash
   C. magnesium
   D. fluoride

10. _____ *True or False?* Superchlorination involves the addition of large doses of chlorine to the water.

11. The _____ scale is used to quantify the level of acidity or alkalinity.

12. _____ On the pH scale, a reading of 0 indicates the water is _____.
    A. extremely acidic
    B. neutral
    C. extremely alkaline
    D. moderately alkaline

Match the parts of the well with the proper part names for questions 13–18.

13. _____ Soda ash chlorine tank
14. _____ Well casing
15. _____ Chemical feeder tube
16. _____ Outlet near pump intake
17. _____ Pump motor
18. _____ Chemical feeder

19. _____ The treatment of water that is moderately acidic or higher may require the use of _____.
    A. sodium hydroxide
    B. soda ash
    C. chlorine
    D. Both A and B.

20. _____ The most trouble-free method of removing iron from water is with a(n) _____ filter charged with potassium permanganate.
    A. oxidizing
    B. activated carbon
    C. neutralized carbon
    D. soda ash

21. _____ *True or False?* Hydrogen sulfide is flammable but nontoxic.

22. _____ _____ filters are cartridge filters that remove calcium and magnesium that produce deposits on piping, fixtures, and accessories.
    A. Sediment
    B. Scale
    C. Activated carbon
    D. Oxidizing

23. _____ Activated carbon (charcoal) filters are primarily used to remove _____.
    A. organic chemicals
    B. chlorine
    C. gases
    D. All of the above.

46 Modern Plumbing Lab Workbook

Name _____

24. _____ A water _____ removes the dissolved calcium and magnesium by ion exchange.
   A. filter
   B. softener
   C. heater
   D. strainer

25. _____ The zeolite softening system process adds _____ to the water.
   A. sodium
   B. magnesium
   C. potassium
   D. Both A and C.

26. _____ *True or False?* The reverse osmosis filtering process is the fastest water treatment process and wastes the least amount of water.

27. _____ _____ is accomplished by heating the water above 160°F (71°C) and retaining it at temperature for an extended period of time.
   A. Reverse osmosis
   B. UV disinfecting
   C. Pasteurization
   D. Boiling

28. _____ *True or False?* Bacteria can be effectively killed by ultraviolet light.

29. _____ Because small particles in the water can shield bacteria from UV light, a(n) _____ filter is necessary.
   A. oxidizing
   B. activated carbon
   C. neutralized carbon
   D. soda ash

30. _____ *True or False?* Plumbers must rely on directions provided by the manufacturer, the health department, and the local plumbing code when installing water treatment systems.

Notes

Name _____ Date _____ Class _____

CHAPTER 15
Plumbing Fixtures

OBJECTIVE: You will be able to recognize the materials from which fixtures are manufactured, identify various fixture types, and distinguish between the different types of toilets (water closets).

Carefully study the chapter and then answer the following questions.

1. _____ _____ is a ceramic material made of kaolin, quartz, feldspar, and silica.
   A. Stainless steel
   B. Acrylic plastic
   C. Porcelain
   D. Gray iron

2. _____ *True or False?* All plumbing fixtures are made of a single material because combinations of materials are prohibited.

3. _____ *True or False?* Porcelain produces a sanitary, easily cleaned surface.

4. _____ Porcelain enamel is sometimes referred to as _____.
   A. "glass lining"
   B. kaolin
   C. vitreous enamel
   D. Both A and C.

5. _____ _____ fixtures are superior to marble fixtures because they do not absorb water.
   A. Stainless steel
   B. Porcelain
   C. Plastic
   D. Cast-iron with porcelain enamel

6. _____ _____ are used for washing hands and faces.
   A. Lavatories
   B. Bidets
   C. Sinks
   D. None of the above.

7. _____ _____ are plumbing fixtures used for food preparation and dishwashing.
   A. Lavatories
   B. Sinks
   C. Faucets
   D. None of the above.

Copyright Goodheart-Willcox Co., Inc.
May not be reproduced or posted to a publicly accessible website.

Match the lavatories and sinks with their proper names.

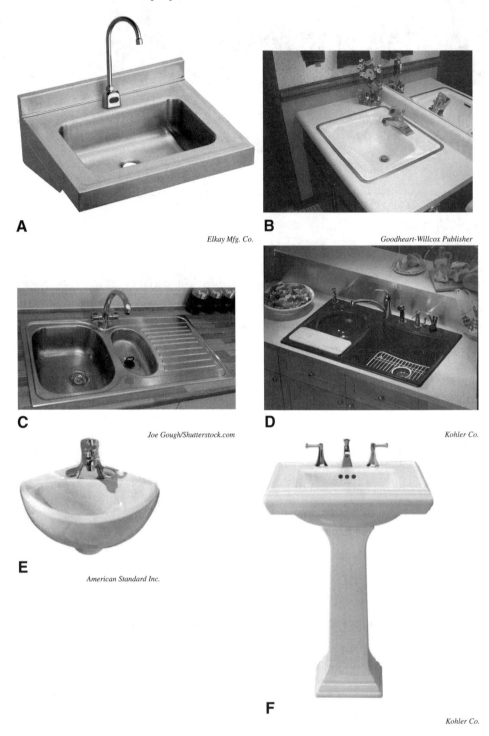

8. _____ One-piece molded type
9. _____ Ledge type
10. _____ Self-rimming type
11. _____ Built-in with metal rim
12. _____ Pedestal type
13. _____ Corner type

Name _____

14. _____ Touch-activated faucets are typically installed in _____ sinks.
    A. basement
    B. kitchen
    C. bathroom
    D. utility

15. _____ Which of the following is *not* a common material used for bathtubs?
    A. Enameled cast iron.
    B. Enameled steel.
    C. Ceramic tile.
    D. Fiberglass-reinforced plastic.

16. _____ A slope of _____″ per foot to the drain outlet provides proper sanitation and reduces maintenance needed on bathtubs.
    A. 1/8
    B. 1-1/4
    C. 1-1/2
    D. 2

17. _____ Plumbers may be expected to install _____ in tubs and similar fixtures for individuals with physical disabilities.
    A. lift devices
    B. special grab bars
    C. seats
    D. All of the above.

18. _____ *True or False?* Scald-proof faucets are installed to prevent the user from being burned.

19. _____ *True or False?* Scald-proof faucets can only control the temperature of the water by manually adjusting a valve.

20. _____ _____ bathtubs pump water and air through jets on the inside of the tub.
    A. Whirlpool
    B. Built-in
    C. Freestanding
    D. Combination

21. _____ The three common sizes of shower bases are 30″ × 30″, 36″ × 36″, and _____.
    A. 36″ × 54″
    B. 36″ × 48″
    C. 48″ × 48″
    D. None of the above.

22. _____ *True or False?* A freestanding shower stall is generally installed in one piece.

23. _____ Shower stall floors are sloped _____″ per foot to a center drain.
    A. 1/8
    B. 1/4
    C. 1/2
    D. 2

24. _____ Toilets are manufactured from _____ or reinforced plastic.
    A. cast iron
    B. vitreous china
    C. polyglycol
    D. ABS-ACR

25. _____ A _____ is designed to carry away solid organic waste using water under pressure or gravity.
    A. urinal
    B. bidet
    C. toilet
    D. garbage disposal

Match each toilet to its proper name for questions 26–30.

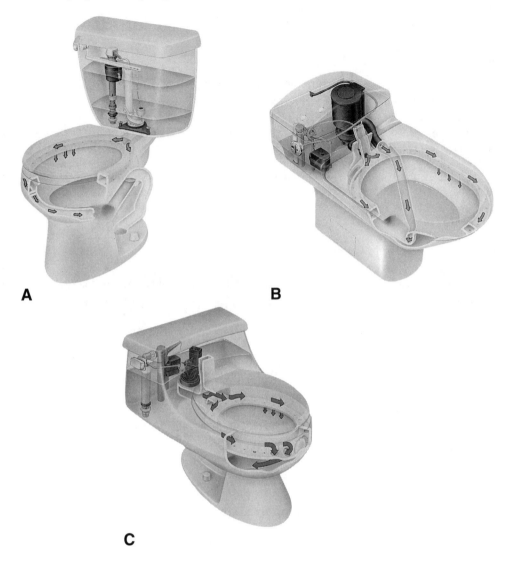

Kohler Co.

26. _____ Pressure-assisted toilet

27. _____ Gravity-fed, rim-jet toilet

28. _____ Gravity-fed, siphon-jet toilet

29. _____ *True or False?* Floor-mounted toilets are designed primarily for commercial buildings.

30. _____ The older traditional toilets use about _____ gallons or more per flush.
    A. 1.6
    B. 3.5
    C. 5
    D. 7

31. _____ The maximum amount of water a toilet can now consume is _____ gallon(s) per flush.
    A. 0.8
    B. 1.6
    C. 2
    D. 3.5

Name _____ Date _____ Class _____

32. _____ The _____ is a companion fixture to the toilet and is used for cleaning the perineal area of the body.
   A. bidet
   B. pressure-assisted toilet
   C. urinal
   D. service sink

33. _____ *True or False?* Urinals are commonly installed in public restrooms for men.

34. _____ The trap of a _____ urinal permits urine to enter a small reservoir through a layer of specially formulated oil that floats on top.
   A. flushless
   B. wall-hung
   C. waterless
   D. None of the above.

35. _____ _____ sinks are found in custodial areas and other locations where cleaning equipment is serviced.
   A. Utility
   B. Service
   C. Pedestal
   D. Self-rimmed

36. _____ *True or False?* Water fountains used in public buildings are required to be wheelchair accessible.

37. _____ The EPA sponsors the WaterSense certification program, which recognizes plumbing products that reduce water consumption by at least _____%.
   A. 5
   B. 10
   C. 20
   D. 50

Notes

Name _____ Date _____ Class _____

CHAPTER 16
Piping Materials and Fittings

OBJECTIVE: You will be able to identify various pipes and fittings and describe their applications, grades, and sizes.

Carefully study the chapter and then answer the following questions.

1. _____ Each plumbing system has _____ set(s) of piping.
    A. one
    B. two
    C. four
    D. six

2. _____ The water supply system carries fresh water under pressure for _____.
    A. drinking
    B. cooking
    C. bathing
    D. All of the above.

3. _____ The _____ system is one set of piping that carries away wastewater and solid waste.
    A. PVC
    B. DWV
    C. ASTM
    D. vent

4. _____ Water _____ refers to the pipe and fittings used to connect the structure with the water main.
    A. supply
    B. service
    C. distribution
    D. All of the above.

5. _____ *True or False?* NSF-61 approval is *not* required for DWV piping.

6. _____ Which of the following information is usually printed along the length of a pipe?
    A. Nominal size.
    B. Type of material.
    C. Standard met.
    D. All of the above.

7. _____ *True or False?* Schedule 40 was a standard originally developed for copper and plastic pipe and fittings.

8. Identify the abbreviated types of plastic.

   A. ABS _____

   B. PVC _____

   C. CPVC _____

   D. PE _____

   E. PEX _____

   F. PP _____

Match each pipe type with its primary application for questions 9–14.

9. _____ ABS
10. _____ PEX
11. _____ CPVC
12. _____ PVC
13. _____ PP
14. _____ PE

A. Hot and cold water supply
B. Hot water and some chemicals
C. Drain, waste, and vent
D. Pressure pipe applications
E. Water service, distribution, and building sewer piping
F. Natural gas

15. _____ Plastic pipe is very popular due to it being _____, competitively priced, easy to cut and join, and having a long service life.
    A. free of chemicals
    B. lightweight
    C. stronger than steel
    D. None of the above.

16. _____ *True or False?* PVC pipe is suitable for hot water piping.

17. _____ *True or False?* PEX is more likely to be damaged by freezing than other commonly used plastic materials.

18. _____ _____ is manufactured for distributing cold and/or hot water inside a structure.
    A. Schedule 40 CPVC
    B. Schedule 80 CPVC
    C. PEX
    D. All of the above.

19. Identify the following abbreviations.

    A. OD _____

    B. ID _____

    C. WT _____

20. _____ *True or False?* Closet flanges connect the water closet to the waste piping.

21. _____ Steel pipe is used for _____.
    A. water service and distribution piping
    B. gas and compressed air piping
    C. aboveground drainage and vent piping
    D. All of the above.

22. _____ The standard length of steel pipe is _____'.
    A. 5
    B. 12
    C. 21
    D. 36

23. _____ *True or False?* Pressure fittings are suitable for drainage systems.

24. _____ The _____ is a fitting that is *not* used to change direction.
    A. elbow
    B. coupling
    C. tee
    D. drop elbow

25. _____ Three grades of steel pipe are standard, _____, and double extra strong.
    A. strong
    B. extra strong
    C. galvanized
    D. None of the above.

Name _____

26. _____ The threads on iron pipe fittings are _____ to form a watertight connection.
    A. tapered
    B. externally threaded
    C. straight
    D. galvanized

Match each type of DWV fitting with the appropriate description.

27. _____ Traps
28. _____ Couplings
29. _____ Elbows
30. _____ Reducers
31. _____ Tees
32. _____ Wyes
33. _____ TYs
34. _____ Plugs
35. _____ Adapters

A. Used to join two lengths of pipe of the same diameter.
B. Used to connect DWV piping systems to other types of DWV piping materials.
C. Couplings that join pipes of two different sizes.
D. Used to close threaded openings.
E. Create a water seal that prevents sewer gases from entering a building.
F. Enable pipe to turn corners.
G. Used to improve flow of waste around corners, similar to sanitary tees.
H. Fittings that create 90° branches to provide access to inside of pipe.
I. Allow a branch to join a run of pipe at a 45° angle.

36. _____ The weight of copper water pipe refers to the _____ of the wall.
    A. height
    B. area
    C. thickness
    D. material

37. _____ The outside diameter of copper pipe is _____″ larger than the standard designation.
    A. 1/32
    B. 1/16
    C. 1/8
    D. 3/16

38. _____ The two types of solder fittings for copper pipe are _____ copper or cast copper.
    A. hard temper
    B. quick-connect
    C. compressed
    D. wrought

39. _____ *True or False?* Dielectric unions can prevent galvanic corrosion on pipes and fittings.

40. _____ *True or False?* Copper DWV fittings are available with inlets ranging from 1-1/2″ to 4″ in diameter.

41. _____ Quick-connect fittings _____.
    A. are available in plastic, brass, and stainless steel
    B. require special tools to install
    C. can be used to join PEX pipe up to 1″ in diameter
    D. Both A and C.

Match the fittings with the proper names for questions 42–45.

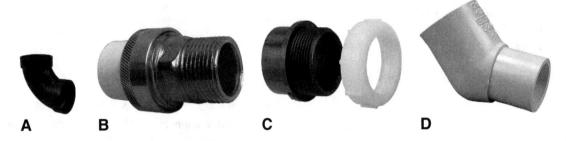

NIBCO, Inc.

42. _____ Union

43. _____ 45° street ell

44. _____ Adapter

45. _____ 90° ell

46. _____ A(n) _____ is a piece of pipe, 12″ or less in length, that is threaded on both ends.
    A. nipple
    B. dielectric union
    C. quick-connect fitting
    D. inverted flare

47. _____ The primary use of vitrified clay pipe is for the installation of _____.
    A. storm sewer mains
    B. sanitary sewer mains
    C. laboratory floor drains
    D. water softeners

48. _____ _____ fittings make a watertight seal when a brass sleeve is squeezed between the nut, pipe, and the body of the fitting.
    A. Inverted flare
    B. Union
    C. Compression
    D. Cap

Name _____ Date _____ Class _____

CHAPTER 17 — Valves and Meters

OBJECTIVE: You will be able to identify and list the types, applications, and construction of various valves, faucets, and meters.

Carefully study the chapter and then answer the following questions.

1. _____ The six materials that valves are generally made of are cast bronze, brass, _____, cast iron, thermoset plastic, and thermoplastic.
    A. polyvinyl chloride
    B. copper
    C. malleable iron
    D. aluminum

2. _____ Valves are available in standard sizes ranging from _____" to _____" diameter.
    A. 1/4, 8
    B. 1/4, 12
    C. 3/4, 8
    D. 3/4, 12

3. _____ *True or False?* Valves may *not* be necessary in a DWV piping system.

4. _____ *True or False?* The stop valve design has remained unchanged for nearly 100 years.

5. _____ Three-way ground key valves are used in _____.
    A. water pumps
    B. gas burner units
    C. automobile washing equipment
    D. All of the above.

6. _____ *True or False?* Corporation valves are available in 2" through 3" diameters.

7. _____ A meter stop allows the water to be shut off at the inlet to the _____.
    A. valve
    B. corporation valve
    C. meter
    D. building

Match valve parts with the proper part name for questions 8–17.

8. _____ Inlet
9. _____ Outlet
10. _____ Valve seat
11. _____ Washer
12. _____ Handle
13. _____ Packing nut
14. _____ Packing box
15. _____ Bonnet
16. _____ Stem
17. _____ Screw thread

Valve

William Powell Co.

18. _____ A _____ valve allows water on the downstream side to be drained.
    A. stop and waste
    B. gate
    C. globe
    D. ball

19. _____ A _____ valve gets its name from the gate-like disk that slides across the path of the flow.
    A. disk
    B. gate
    C. globe
    D. ball

20. _____ A _____ valve is open with just one-quarter turn and provides little resistance to flow.
    A. gate
    B. ball
    C. flush
    D. float-controlled

21. _____ *True or False?* Toilets or urinals that use flush valves do *not* need a storage tank.

22. _____ *True or False?* Gravity flow in a storage tank generally cleans better than a flush valve.

23. _____ _____ flush valves use infrared light beams to detect when a fixture has been used to activate flushing.
    A. Manual
    B. Diaphragm
    C. Sensor-activated
    D. Dual

24. _____ *True or False?* A control stop is generally installed in the water supply line serving the flush valve.

25. _____ A _____ is installed between the outlet of the flush valve and fixture to prevent back siphoning.
    A. backflow preventer
    B. diaphragm
    C. vacuum breaker
    D. float-controlled valve

26. _____ *True or False?* Backflow preventer plumbing devices include the double-check valve, reduced pressure zone valve, and the spill-resistant vacuum breaker assembly.

27. _____ A _____ valve allows excess pressure to bleed off to the atmosphere.
    A. float
    B. temperature/pressure relief
    C. pressure regulator
    D. back water

28. Identify the parts of the float-controlled valve.

    A. _____

    B. _____

    C. _____

    Fluidmaster, Inc.

Name _____

29. _____ A(n) _____ reduces water pressure in a building.
    A. stop valve
    B. angle valve
    C. pressure regulator
    D. relief valve

30. _____ A pressure regulator valve is installed if the water pressure exceeds _____.
    A. 40 psi
    B. 60 psi
    C. 80 psi
    D. 100 psi

31. _____ *True or False?* Back water valves are a type of check valve designed for DWV piping systems.

32. _____ A _____ is a mechanical device that measures the amount of water passing through the water service pipe into a building.
    A. leak detection and control device
    B. water meter
    C. water detection device
    D. None of the above.

33. _____ The water meter is usually the property of the _____.
    A. landowner
    B. city
    C. plumbing contractor
    D. None of the above.

34. _____ Water meters are manufactured with capacities of _____ gallons per minute.
    A. 15, 20, or 25
    B. 20, 25, or 30
    C. 20, 30, or 50
    D. 25, 35, or 45

Notes

Name _____ Date _____ Class _____

CHAPTER 18
Water Heaters

OBJECTIVE: You will be able to identify the two types of heating/storage tanks, describe their differences, demonstrate how their controls work, and explain the steps for water heater installation.

Carefully study the chapter and then answer the following questions.

1. _____ The _____ is responsible for installing the piping for water heaters.
    A. plumber
    B. electrician
    C. pipe fitter
    D. HVAC contractor

2. _____ *True or False?* Storage-type water heaters are the most common water heating equipment.

Match the operational parts and processes of the water heater with the proper names for questions 3–11.

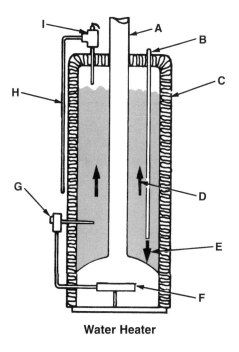

Water Heater

Goodheart-Willcox Publisher

3. _____ Thermostat

4. _____ Temperature/pressure relief valve

5. _____ Flue

6. _____ Cold water inlet

7. _____ Insulation

8. _____ Hot water rises

9. _____ Heating unit

10. _____ Cold water sinks to the bottom of tank

11. _____ Overflow piping

12. _____ In a storage-type water heater, the cold water enters the tank near the _____.
    A. center
    B. top
    C. bottom
    D. All of the above.

13. _____ In a storage-type water heater, hot water is drawn off near the _____ of the tank.
    A. top
    B. bottom
    C. center
    D. None of the above.

14. _____ When water is heated, it _____.
    A. contracts
    B. expands
    C. condenses
    D. None of the above.

15. _____ _____ is the most common energy source for water heaters.
    A. Natural gas
    B. Electricity
    C. Oxygen
    D. Carbon

16. _____ *True or False?* Natural gas is much more concentrated than liquefied petroleum gas (LPG).

17. _____ *True or False?* Electric water heaters use an insulated heating element to heat the water.

18. _____ Most electric water heaters require _____V wiring.
    A. 120
    B. 220
    C. 240
    D. None of the above.

19. _____ *True or False?* In normal operation of an electric heater, only the top coil is working.

20. In heat pump water heaters (HPWH), refrigerant is pumped through a _____ to increase the temperature of the refrigerant.
    A. heating coil
    B. condensate pump
    C. heat exchanger
    D. combustion chamber

21. _____ The _____ is the speed at which cold water can be heated.
    A. heated rate
    B. recovery rate
    C. energy efficiency
    D. set point

22. _____ When selecting a storage-type water heater, the _____ should be considered.
    A. capacity of the water heater
    B. fuels available
    C. ability of the tank to hold heat
    D. All of the above.

23. _____ The Energy Guide Program is administered by the _____.
    A. Federal Trade Commission
    B. American Gas Association
    C. American Water Worker
    D. Underwriters Lab

64 Modern Plumbing Lab Workbook

Name _____

24. _____ A(n) _____ is an automatic device that turns the energy source of the water heater on and off as needed.
    A. set point
    B. thermostat
    C. standing pilot
    D. ignition system

Match the parts of the valve control unit with the proper part names for questions 25–34.

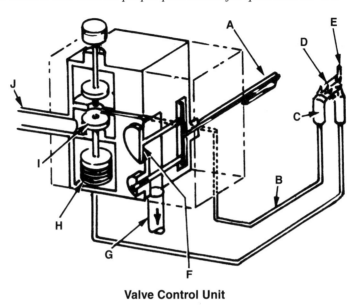

Valve Control Unit

Goodheart-Willcox Publisher

25. _____ Gas supply line

26. _____ Solenoid valve

27. _____ Solenoid coil

28. _____ Heat-sensing element

29. _____ Pilot

30. _____ Pilot supply line

31. _____ Thermocouple

32. _____ Burner supply line

33. _____ Thermostat valve

34. _____ Flame

35. _____ The _____ cuts off the current to the solenoid should the water temperature exceed a predetermined safe limit.
    A. standing pilot
    B. safety cutoff thermostat
    C. thermostat valve
    D. T/P relief valve

36. _____ The high-limit protector is an automatic safety device that shuts off all electrical current if the water temperature exceeds _____.
    A. 180°F
    B. 180°C
    C. 82°C
    D. Both A and C.

37. _____ *True or False?* The temperature/pressure (T/P) relief valve is installed in a specially designed opening in the top of the tank, and prevents excess temperature and pressure buildup in the tank.

38. _____ The major hazard of water heaters is related to the buildup of _____ water in the tank.
   A. superheated
   B. warm
   C. cold
   D. potable

39. _____ When water changes to steam, its volume increases approximately _____ times.
   A. 100
   B. 800
   C. 1600
   D. 2400

40. _____ If water does not flow out of the drip line when the T/P relief valve is manually opened to check it, then _____ the valve immediately.
   A. adjust
   B. replace
   C. protect
   D. close

41. _____ *True or False?* Gas-fired water heaters should *not* be placed near a chimney or flue.

42. _____ *True or False?* Directly connecting a gas-fired water heater to a furnace flue is often permitted if a T connector is used.

43. _____ For a higher volume of hot water, two or more water heaters may be connected in _____.
   A. parallel
   B. series
   C. series-parallel
   D. All of the above.

44. _____ When using a solar collector to heat water, it is typical to install _____.
   A. two water heaters in parallel
   B. a smaller water heater
   C. two water heaters in series
   D. None of the above.

45. _____ A(n) _____ water heater is most frequently used in the United States for point-of-use heating of water.
   A. electric
   B. gas-fired
   C. instantaneous
   D. solar

46. _____ In order to correctly size an instantaneous hot water heater, accurate estimates of _____ must be made.
   A. number of gallons used per minute
   B. flow rates
   C. water supply piping
   D. hot water demand

47. _____ An important consideration with instantaneous water heaters is _____.
   A. energy source
   B. installation cost
   C. low flow rate
   D. Both B and C.

Name _____

Match each of these common fixtures with its hot water demand for questions 48–53.

48. _____ Kitchen sink
49. _____ Dishwasher
50. _____ Bathtub
51. _____ Shower
52. _____ Lavatory
53. _____ Washing machine

A. 3.6 GPM
B. 1.5 GPM
C. 1.6 GPM
D. 0.3 GPM
E. 2.5 GPM
F. 3.3 GPM

54. _____ *True or False?* Gas- or oil-fired instantaneous units require exterior venting.

55. _____ *True or False?* Heating water with solar collectors does little to reduce the energy needed to heat water.

56. _____ _____ extract heat from the outside air and transfer it to the water.
    A. Insulating pipes
    B. Heat pumps
    C. Heat extractors
    D. Solar collectors

57. _____ Hot water pipes from the water heater to each fixture should be _____ to reduce energy consumption.
    A. sealed tight
    B. solar-assisted
    C. insulated
    D. limited

Notes

Name _____ Date _____ Class _____

CHAPTER 19
Designing Plumbing Systems

OBJECTIVE: You will be able to determine the size of drainage and water supply piping, design efficient and serviceable plumbing systems, and plan plumbing installations to serve disabled individuals. You will be able to recognize and avoid cross connections and describe storm water collection and drainage systems.

Carefully study the chapter and then answer the following questions.

1. _____ *True or False?* When designing a plumbing system, only the architect can identify basic needs for the building.

2. _____ *True or False?* In residential buildings, the number of people who use the restroom facilities must be considered.

3. _____ The _____ of 1990 requires places of employment and structures used by the general public be accessible to physically disabled individuals.
   A. Standards for Accessible Design
   B. Americans with Disabilities Act (ADA)
   C. Accessible and Usable Buildings and Facilities Code (AUBF)
   D. Fair Housing Accessibility Guidelines

4. _____ *True or False?* To reduce costs, rooms needing plumbing should be placed near each other.

5. _____ It is common practice to arrange rooms needing plumbing _____.
   A. back-to-back
   B. above and below each other
   C. at all corners of the building
   D. Both A and B.

6. _____ *True or False?* It is *not* good practice to have one soil stack serving more than one room.

7. _____ Deciding the placement of rooms is done by the _____.
   A. plumber
   B. owner
   C. architect
   D. AUBF

8. _____ ADA requirements specify the clear floor space around a lavatory, sink, or urinal must be a minimum of _____" by _____".
   A. 30; 48
   B. 48; 66
   C. 50; 56
   D. 60; 36

9. _____ *True or False?* The ADA requires at least one handicapped accessible restroom on each floor.

10. _____ While planning a kitchen, a planner must be concerned with the work area "triangle" created by the _____, refrigerator, and sink.
    A. garbage disposal
    B. hot and cold water piping
    C. stove
    D. dishwasher

11. _____ *True or False?* Sinks, dishwashers, and garbage disposals require DWV piping.

12. _____ The _____ dimensions indicate where the DWV and water supply must be located in the wall and/or floor.
    A. utility
    B. accessibility
    C. ADA
    D. rough-in

13. _____ The _____ pipe system operates on gravity.
    A. sanitary water supply
    B. DWV
    C. water distribution
    D. All of the above.

14. _____ A drop of _____" to _____" per foot is considered adequate depending on the diameter of the pipe in a DWV system.
    A. 1/8; 1/4
    B. 1/4; 1/2
    C. 1/2; 3/4
    D. 3/4; 2

15. _____ A DWV pipe that is too large is _____.
    A. likely to clog
    B. extremely expensive
    C. more difficult to install
    D. All of the above.

16. _____ There are approximately _____ gallons of water in one cubic foot.
    A. 7.5
    B. 9
    C. 3.25
    D. 8.5

17. _____ A lavatory that discharges approximately _____ gallon(s) of water per minute has a load factor of 3.
    A. 1
    B. 7.5
    C. 15
    D. 22.5

18. _____ _____ indicate the load each fixture contributes to what the pipe must be able to accommodate.
    A. Load factors
    B. Drainage fixture units
    C. Stacks
    D. None of the above.

19. *True or False?* Stacks must never be smaller than the largest branch entering them.

20. _____ _____ permit air to circulate through the waste piping system.
    A. Stacks
    B. Drains
    C. Vents
    D. Supply pipes

21. _____ A _____ in the waste line prevents sewer gas from entering the building by providing a water seal.
    A. trap
    B. cleanout
    C. vent
    D. None of the above.

Name _____

Match the ways a trap can lose its water seal with its description for questions 22–26.

22. _____ Indirect siphonage
23. _____ Back pressure
24. _____ Evaporation
25. _____ Capillary action
26. _____ Wind

A. Downdraft in stack
B. Foreign material caught in trap
C. System not in use for extended period
D. Water drawn from one fixture by discharge of another
E. Air pressure builds in system

27. _____ _____ venting is the best venting method because it vents every trap separately.
    A. Individual
    B. Unit
    C. Circuit
    D. Relief

28. _____ *True or False?* The fittings used to install vent piping are the same as those used for the waste piping.

29. _____ The water supply system is designed to operate between _____ psi within the building.
    A. 10–20
    B. 30–40
    C. 40–60
    D. 80–100

30. _____ _____ pipe is *not* suitable for installation of the water supply system.
    A. Black iron
    B. CPVC
    C. Lead
    D. Both A and C.

31. _____ In a water tower, the pressure increases _____ psi for each foot of height.
    A. 0.433
    B. 1
    C. 10
    D. 50

Using the tower diagram, find the water pressure for each floor for questions 32–38.

32. _____ Basement
33. _____ 1st floor
34. _____ 2nd floor
35. _____ 3rd floor
36. _____ 4th floor
37. _____ 5th floor
38. _____ 6th floor

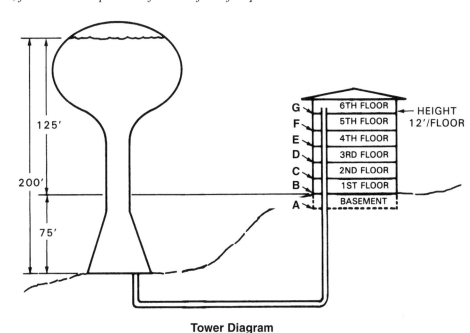

Tower Diagram

Goodheart-Willcox Publisher

39. _____ *True or False?* Consideration must be given for pressure loss through piping, even on small piping jobs.

40. _____ Water piping must be protected from temperatures below 32°F (0°C) because the pipe may _____.
    A. freeze
    B. burst
    C. be damaged
    D. All of the above.

41. _____ *True or False?* The number of fittings installed has no effect on water pressure.

42. _____ *True or False?* Air gap applications are used to prevent cross connections.

43. _____ A(n) _____ valve is used to prevent a water heater from exploding if the thermostat ceases to function.
    A. shutoff
    B. pressure-relief
    C. venting
    D. air gap

44. _____ Shutoff valves allow the plumber to _____ small parts of the piping system.
    A. clean
    B. isolate
    C. supply
    D. None of the above.

45. _____ *True or False?* Piping systems containing potable water can be interconnected with other piping systems.

46. _____ _____ is caused by the sudden stop of water flow.
    A. Water heater failure
    B. Vibration
    C. Water hammer
    D. Loss of water pressure

47. _____ _____ are installed in water piping to prevent water hammer.
    A. Relief valves
    B. Air chambers
    C. Bypass valves
    D. Pressure-reducing valves

Name _____ Date _____ Class _____

CHAPTER 20
Preparing for Plumbing System Installation

OBJECTIVE: You will be able to describe steps involved in locating and bringing in DWV and water supply in a building. You will be able to use the prescribed techniques for working with plumbing materials and fixtures and describe methods of laying out plumbing systems.

Carefully study the chapter and then answer the following questions.

1. Using the numbers 1, 2, 3, and 4, place stages for installing plumbing in proper order.
    A. _____ Sink, lavatory, and showerhead installation
    B. _____ Piping enclosed in the walls
    C. _____ Sewer and water supply installation
    D. _____ Final inspection

2. _____ _____ are required to make connections to the water main and the sanitary sewer.
    A. Mechanical tees
    B. Adapters
    C. Permits
    D. None of the above.

3. _____ If permitted by local code, water supply and building sewer pipes can be run in a _____ trench that provides a minimum distance between the pipes.
    A. small
    B. stepped
    C. backhoe
    D. large

4. _____ *True or False?* Once the building sewer is installed, it must be inspected before backfilling.

5. _____ The two different methods used to attach the corporation stop to the water main are to drill and tap the side wall of the main or to attach a(n) _____ to the main.
    A. curb box
    B. meter yoke
    C. saddle
    D. angle meter valve

6. _____ The _____ has responsibility for installing the water line from the curb stop into the house.
    A. plumbing contractor
    B. local municipality
    C. architect
    D. None of the above.

7. _____ A(n) _____ is a valve installed near the point where the water supply line enters the building so a water meter can be attached.
    A. angle meter valve
    B. meter yoke
    C. curb box
    D. Both A and B.

8. _____ The _____ is the drainage piping between the building's foundation and the sewer main or septic tank.
    A. curb box
    B. DWV
    C. building sewer
    D. water main

Copyright Goodheart-Willcox Co., Inc.
May not be reproduced or posted to a publicly accessible website.

9. _____ *True or False?* Allowances for finished walls and flooring are an important part of the second rough.

10. _____ *True or False?* A curb box is a control valve installed between the corporation stop and the structure.

11. _____ With cast-in-place concrete floors aboveground, _____ should be used to produce opening for pipes to pass from floor-to-floor.
    A. special fittings
    B. cans
    C. fasteners
    D. additional concrete

12. _____ Rough-in dimensions for fixtures are supplied by the _____.
    A. architect
    B. plumbing contractor
    C. manufacturer
    D. city

13. _____ _____ strips may be needed to make a wall wide enough to conceal piping.
    A. Furring
    B. Sole
    C. Stacking
    D. Wooden

14. _____ When possible, the stack should be installed directly behind the _____.
    A. lavatory
    B. tub
    C. toilet
    D. shower stall

15. _____ The water supply piping is installed after the _____ system is completed.
    A. gas
    B. PVC
    C. ABS
    D. DWV

16. _____ *True or False?* The cold water supply for the toilet will always come up through the floor.

17. _____ *True or False?* The lavatory water supply piping should stub out through the wall.

18. _____ When cutting floor joists, _____ joists and headers ensure that the strength of the floor remains.
    A. notched
    B. doubled-up
    C. tripled-up
    D. stubbed

19. _____ A(n) _____ is installed in the wall to support the weight of a wall-hung lavatory.
    A. 2 × 6 block
    B. cast-iron pipe
    C. extra stud
    D. large nail

20. _____ Commercial buildings often have wall-hung _____.
    A. toilets
    B. urinals
    C. lavatories
    D. All the above.

21. _____ After all _____ have been rechecked for accuracy, the drilling and cutting of openings can begin.
    A. fixtures
    B. layout dimensions
    C. stack sizing
    D. water supply piping

Name _____ Date _____ Class _____

CHAPTER 21
DWV Pipe and Fitting Installation

OBJECTIVE: You will be able to describe procedures for locating and installing DWV piping in a building. You will be able to use the prescribed techniques for working with plumbing materials, use three methods of measuring pipe, and describe methods of testing and inspecting plumbing systems.

Carefully study the chapter and then answer the following questions.

1. _____ The building sewer and water supply are installed during the _____.
   A. first rough
   B. second rough
   C. third rough
   D. final inspection

2. _____ Which of the following is *not* one of the three types of measurements taken to determine length of pipe?
   A. Face-to-face.
   B. Shoulder-to-shoulder.
   C. Toe-to-toe.
   D. Center-to-center.

3. _____ The _____ method of determining length of pipe is useful only with DWV fittings.
   A. face-to-face
   B. shoulder-to-shoulder
   C. toe-to-toe
   D. center-to-center

4. _____ The face-to-face method requires that the plumber know the fitting _____ for the type and size of fitting being used.
   A. allowance
   B. laying length
   C. load
   D. material

5. _____ The _____ is the amount of a run of a pipe that is taken up by the fitting.
   A. allowance
   B. fixture load
   C. laying length
   D. diameter

6. _____ During DWV installation, it is better for the _____ to be slightly higher than necessary so the cleanout plug can be easily installed and removed.
   A. tee
   B. wye
   C. stack
   D. plumb bob

7. _____ A _____ will allow the dimensions to be checked before final assembly.
   A. tee
   B. trial assembly
   C. standpipe
   D. None of the above.

8. _____ *True or False?* The horizontal pipes to the standpipe and laundry tub should *never* be installed after the concrete floor has been poured.

Copyright Goodheart-Willcox Co., Inc.
May not be reproduced or posted to a publicly accessible website.

75

9. _____ *True or False?* The amount of the stack that extends above the roof is specified by code.

10. _____ In cold climates, larger pipes are used above the insulated area of the building to prevent the _____ from becoming blocked by frozen condensation.
    A. chimney
    B. stack
    C. vent
    D. coupling

11. _____ _____ is installed on the vent pipe at the roof line to prevent roof leaks around the pipe.
    A. Flashing
    B. Insulation
    C. Extra piping
    D. Special roofing material

12. _____ The main concern when installing horizontal runs of drain/waste piping is maintaining the required _____ for the entire length of the run.
    A. water pressure
    B. fall
    C. flashing
    D. Both A and C.

13. _____ Bathtubs and showers are installed as recessed, _____, corner, and peninsular.
    A. standing
    B. one-piece molded
    C. sunken
    D. assembled

14. _____ *True or False?* One-piece molded plastic tub/shower units are typically installed after the framing for the bathroom walls is completed.

15. _____ *True or False?* DWV plastic piping is generally measured using the direct and indirect methods.

16. _____ A fine-tooth saw should be used with a _____ when cutting PVC pipe to ensure that the cut is square.
    A. pipe cutter
    B. miter box
    C. plastic hanger
    D. tubing cutter

17. _____ Horizontal runs of plastic pipe should be supported every _____′ to _____′.
    A. 2; 4
    B. 3; 4
    C. 4; 6
    D. 6; 8

18. _____ *True or False?* Copper DWV pipe measurements are generally taken center-to-center.

19. _____ *True or False?* When cutting copper pipe with a tubing cutter, the cut end of the pipe should be reamed.

20. _____ Horizontal runs of copper pipe should be supported every _____′ to _____′.
    A. 2; 4
    B. 3; 4
    C. 4; 6
    D. 6; 8

21. _____ Copper DWV pipe and fittings are joined by _____.
    A. solvent
    B. brazing
    C. soldering
    D. welding

22. _____ *True or False?* Black iron pipe is used for DWV piping.

Name _____

23. _____ When threading pipe, _____ should be used to lubricate the die.
    A. solvent cement
    B. flux
    C. warm water
    D. cutting oil

24. _____ _____ or pipe joint sealer should be applied to the pipe threads before joining the pipe and fitting.
    A. Teflon tape
    B. Flux
    C. Pipe adhesive
    D. Stainless steel clamps

25. _____ Galvanized and black steel pipe should be supported at _____′ intervals.
    A. 2
    B. 4
    C. 12
    D. 24

26. _____ *True or False?* Cast-iron pipe is generally cut with a compound lever or hydraulic pipe cutter.

27. _____ No-hub cast-iron pipe is joined with a _____ and a stainless steel clamp.
    A. pipe joint sealer
    B. neoprene gasket
    C. lead and oakum seal
    D. silicone lubricant

28. _____ Vertical runs of cast-iron pipe can be attached to the building structure with vertical pipe brackets or _____.
    A. Teflon tape
    B. lead and oakum
    C. pipe straps
    D. neoprene gaskets

Notes

Name _____ Date _____ Class _____

CHAPTER 22
Installing Water Supply Piping

OBJECTIVE: You will be able to explain the methods of measuring pipe between fittings, describe proper procedures for locating and installing water supply systems, and use prescribed techniques for working with pipes and fittings made from PVC, CPVC, PEX, and copper.

Carefully study the chapter and then answer the following questions.

1. _____ The second rough stage of plumbing installation includes _____ covered by finished wall material.
   A. DWV piping
   B. water supply piping
   C. Both A and B.
   D. None of the above.

2. _____ *True or False?* Water pressure at the meter can be determined by the local water authority.

3. _____ *True or False?* The difference in elevation of the water supply pipe at the meter and the highest and lowest water supply outlet valves will *not* affect the water pressure at the valve.

4. _____ *True or False?* The maximum developed length from the meter is determined at the same time for both hot and cold water piping.

5. List the flow rates for the following fixtures:

   A. Lavatory faucet _____

   B. Showerhead _____

   C. Sink faucet _____

   D. Toilet _____

6. _____ The water filter is installed in the _____ from the water meter.
   A. meter yoke
   B. pipe run
   C. floor joists
   D. None of the above.

7. _____ The IRC does not require a _____ to be installed on the inlet side of a water filter.
   A. hose bibb
   B. cartridge
   C. full-open valve
   D. All of the above.

8. _____ The water meter is generally installed at the point where the building water supply pipe extends through the _____.
   A. water main
   B. side of the building
   C. meter yoke
   D. foundation wall

9. _____ The water supply pipe from the meter divides into two main branches at the water heater. One branch is for cold water and the branch that connects to the _____.
   A. DWV piping
   B. dip tube
   C. water heater
   D. T/P relief valve

10. _____ The cold water inlet of a water heater is fitted with a(n) _____ that directs entering cold water to the bottom of the tank.
    A. T/P relief valve
    B. dip tube
    C. PVC fitting
    D. MIPT adapter

11. _____ Before extending hot and cold water lines from the water heater to the kitchen and bath areas, check the position of the _____ in each location to determine the most practical route to run the pipes.
    A. stub-outs
    B. appliances
    C. plumbing fixtures
    D. DWV piping

12. _____ The cold water stub-out is always installed on the _____ side, while the hot stub-out is always on the left side.
    A. inlet
    B. right
    C. front
    D. back

13. _____ Plumbing for a clothes washer is installed inside a finished wall, with the pipes connected to a washing machine _____ box.
    A. inlet
    B. outlet
    C. square
    D. None of the above.

14. _____ Which of the following is *not* one of the four pipe connections for tub/shower valves?
    A. Hot water
    B. Cold water
    C. Toilet
    D. Showerhead

15. _____ Most tub/shower valves require _____ for connecting pipes.
    A. MIPT adapters
    B. drop ells
    C. copper fittings
    D. caps

16. _____ *True or False?* PVC should be used for the connections hot water piping.

17. _____ PVC pipe is white, and CPVC is _____.
    A. black
    B. tan
    C. gray
    D. brown

18. _____ _____ is the change in dimensions of water supply piping materials caused by fluctuating temperature.
    A. Expansion looping
    B. Thermal expansion
    C. Thermal contraction
    D. Thermal fluctuation

19. _____ Assume that a 175′ straight run of CPVC water supply pipe was installed when the temperature was 70°F. If the temperature rises to 95°F, how much longer would the pipe be?
    A. 1″
    B. 1.58″
    C. 1.66″
    D. 16.63″

Name _____

20. _____ Expansion loops should be installed near the _____ of long runs of pipe.
    A. end
    B. beginning
    C. center
    D. All of the above.

21. _____ Most plastic water supply lines are measured using the _____ technique.
    A. face-to-face
    B. shoulder-to-shoulder
    C. center-to-center
    D. back-to-back

22. _____ When joining PVC pipe and fittings with solvent cement, allow _____ hours before pressure testing.
    A. 2
    B. 4
    C. 12
    D. 24

23. _____ The major difference between copper pipe and fitting installation and plastic pipe installation is that the joints are _____ rather than joined with adhesive.
    A. soldered
    B. brazed
    C. welded
    D. taped

24. _____ When it is necessary to use center-to-center measurements for copper pipe and tubing installation, the _____ of copper pressure fittings is needed.
    A. total diameter
    B. laying length
    C. total allowances
    D. number

25. _____ Copper pipe and tubing should be cut with a _____.
    A. hand saw
    B. plastic pipe cutter
    C. tubing cutter
    D. clamp

26. _____ The cut end of a pipe is reamed to remove the _____.
    A. adhesive
    B. solder
    C. pipe spigot
    D. burr

27. _____ *True or False?* Installing PEX differs in several ways from installing rigid pipe such as PVC, CPVC, and copper.

28. _____ PEX joints are made with _____ rings and a special crimping tool.
    A. galvanized
    B. crimp
    C. composite
    D. rigid

29. _____ PEX pipe expands and contracts _____" for every _____' of tube for every 10°F change in temperature.
    A. 1; 100
    B. 2; 200
    C. 3; 300
    D. All of the above.

30. _____ Installing PEX and composite plastic tube is more like installing _____ than plumbing pipe.
    A. venting
    B. electrical wire
    C. appliances
    D. All of the above.

31. _____ _____ are often used with PEX and composites to reduce the need to install varying sizes of pipe and isolate each group of plumbing fixtures.
    A. Special bend supports
    B. Quick-connect fittings
    C. Manifolds
    D. Special brackets

32. List at least six questions that might be asked before routing the water supply piping.

    _____

    _____

    _____

    _____

    _____

    _____

33. List three interrelated factors that the size of the pipe and fittings are dependent upon.

    _____

    _____

    _____

    _____

Name _____ Date _____ Class _____

CHAPTER 23
Supporting and Testing Pipe

OBJECTIVE: You will be able to install appropriate supports for residential and light commercial piping systems, install anchors in concrete and masonry, and test both DWV and water supply piping systems.

Carefully study the chapter and then answer the following questions.

1. _____ *True or False?* Both DWV and water supply piping must be supported both horizontally and vertically.

2. _____ Horizontal supports prevent sagging, maintain alignment of both DWV and water supply piping, and maintain the _____ of the DWV piping.
   A. weight
   B. fall
   C. water pressure
   D. distribution

3. _____ Vertical supports maintain alignment and support the _____ of piping systems.
   A. weight
   B. fall
   C. water pressure
   D. distribution

4. _____ *True or False?* Some of the needed supports are installed as pipe and fittings are put in place.

5. _____ The maximum horizontal spacing of pipe hangers varies from 2′-8″ to _____′.
   A. 4
   B. 10
   C. 12
   D. 16

6. _____ The _____ and weight of the pipe, the material from which it is made, and the position it must be supported in are three important factors in making the selection of pipe hangers and supports.
   A. fall
   B. size
   C. capacity
   D. price

7. _____ Which of the following is *not* typically used to support copper pipe and fittings horizontally?
   A. Plumber's tape.
   B. Wire pipe staples.
   C. Pipe straps.
   D. Pipe clamps.

8. _____ *True or False?* When supporting metal pipe, hangers that permit expansion and contraction movement should be used.

9. _____ Long horizontal runs of DWV piping must be supported in a way that maintains a constant _____.
   A. water pressure
   B. slope
   C. length
   D. expansion

Copyright Goodheart-Willcox Co., Inc.
May not be reproduced or posted to a publicly accessible website.

10. _____ Supporting the weight of DWV vertical pipe is accomplished by securing _____ at each floor level.
    A. pipe straps
    B. U-bolts
    C. riser clamps
    D. prepunched channels

11. _____ A _____ is often used to make pipe supports for parallel pipe runs.
    A. channel
    B. pipe clamp
    C. pipe strap
    D. strut

12. _____ *True or False?* Screws and glue are used to secure pipe hangers and supports to wood-framed structures.

13. _____ *True or False?* In steel-framed buildings, clamp-like devices are often used to attach hangers for smaller pipes.

14. _____ Concrete anchors may use _____ to secure objects.
    A. hex head bolts
    B. machine screws
    C. hex nuts
    D. All of the above.

15. _____ Stud-type anchors allow the _____ to be removed if the anchored object needs to be removed.
    A. bolt
    B. nut
    C. screw
    D. clamp

16. _____ *True or False?* When using anchors that require a large diameter hole, it is best to install the anchor after drilling the holes in the object to be anchored.

Match each anchor with its proper name for questions 17–21.

17. _____ Stud
18. _____ Drop-in anchor
19. _____ Self-drilling
20. _____ Insert
21. _____ Full thread sleeve

Anchors

ITW Ramset/Red Head; Phillips Drill Co.

Name _____

Match the pipe hangers and supports with their proper names for questions 22–29.

22. _____ Pipe clamp
23. _____ I-beam clamp
24. _____ Beam clamp
25. _____ Steel "C" clamp
26. _____ Pipe strap
27. _____ Swivel loop hanger
28. _____ Wall bracket
29. _____ Clevis hanger

Pipe Hangers and Supports

Goodheart-Willcox Publisher; Modern Hanger Corp.; NIBCO, Inc.

30. _____ _____ anchors provide a fastener that is permanently attached to the concrete or masonry and is internally threaded to accept machine screws or bolts.
    A. Caulking
    B. Plastic
    C. Self-tapping
    D. Self-drilling

31. _____ _____ are useful when attaching pipe hangers to hollow masonry units.
    A. Lag shields
    B. Caulking anchors
    C. Toggle bolts
    D. Plastic anchors

32. _____ Masonry bits are available in sizes from 1/8″ to _____″.
    A. 5/32
    B. 1/2
    C. 1-1/2
    D. 2

33. _____ *True or False?* The plumber must contact the plumbing inspector to initiate the rough-in inspection process.

34. _____ _____ are used to seal openings in a piping system in preparation for air and water tests.
    A. Hydraulic test pumps
    B. Test plugs
    C. Hoses with pressure gauges
    D. Soapsuds

35. _____ A _____ may be used to produce pressures as high as 400 psi.
    A. hydraulic test pump
    B. test plug
    C. pneumatic test plug
    D. pressure gauge

36. _____ When air testing a DWV pipe system, a pressure of _____ psi is generally adequate.
    A. 20
    B. 15
    C. 10
    D. 5

Notes

Name _____ Date _____ Class _____

CHAPTER 24
Installing Water Heaters, Fixtures, Faucets, and Appliances

OBJECTIVE: You will be able to describe proper installation procedures for each fixture, faucet, and appliance presented in this chapter. You will also identify the specials tools needed to install the various fixtures and procedures for connection to the water supply and DWV connections.

Carefully study the chapter and then answer the following questions.

1. _____ *True or False?* Due to the unique characteristics of the various fixtures, plumbers should study the manufacturer's instructions before attempting to install any fixture.

2. _____ Tank-type water heater installation is typically delayed until the _____ of plumbing installation.
   A. first-rough
   B. second-rough
   C. finish stage
   D. inspection

3. _____ Plumbing code requires that a valve be installed on the _____ entering the water heater.
   A. hot water pipe
   B. cold water pipe
   C. T/P relief valve
   D. discharge pipe

4. _____ When installing the discharge pipe on a water heater, ensure that the pipe ends near a(n) _____ or extends outside the building.
   A. vent
   B. floor drain
   C. exit of the room
   D. building sewer main

5. _____ _____ pipe and fittings are commonly used to deliver fuel gas inside of buildings.
   A. Black iron
   B. PVC
   C. Cast-iron
   D. CPVC

6. _____ *True or False?* Both electric and gas water heaters must be filled with water *before* the electricity or gas is turned on.

7. _____ Mounting the _____ before installing the lavatory or sink makes it much easier to reach and tighten the nuts on the underside.
   A. rim
   B. faucet
   C. P-trap
   D. cleanout

8. _____ Water supply tubes are made from flexible PVC tubing, braided stainless steel, and _____ soft copper tubing.
   A. galvanized
   B. flexible
   C. chrome-plated
   D. reinforced

9. _____ *True or False?* The adapter fittings for the water supply and drain piping stub-outs should be installed after putting the lavatory or sink in place.

10. _____ Lavatories are typically _____" above the floor.
    A. 12
    B. 30
    C. 36
    D. 40

11. _____ The drain from a _____ lavatory extends from the bowl of the lavatory to where the P-trap is connected.
    A. pedestal-type
    B. self-rimming
    C. wall-mounted
    D. metal rim

12. _____ *True or False?* When self-rimming and metal-rimmed lavatories and sinks are installed, adhesive caulk is used to seal the joint at the countertop.

13. _____ When installing drainage fittings, the _____ is installed after the sink or lavatory is placed in position.
    A. faucet
    B. lug
    C. strainer body
    D. adhesive caulk

14. _____ _____ is used to seal the joint between the strainer body and the sink or lavatory.
    A. Adhesive caulk
    B. Plumber's putty
    C. Electric tape
    D. Flexible tubing

15. _____ *True or False?* A strainer locknut wrench is used to prevent the strainer body from rotating, while a strainer wrench is used to tighten the nut.

16. _____ Kitchen sinks are frequently fitted with a(n) _____.
    A. garbage disposal
    B. ballcock
    C. indirect waste valve
    D. None of the above.

17. _____ The outlet of a garbage disposal is connected directly to the DWV piping with _____ pipe and fittings.
    A. plastic
    B. copper
    C. slip-joint
    D. compression

18. _____ The minimum diameter for the fixture tailpiece of a lavatory is _____.
    A. 1"
    B. 1-1/4"
    C. 1-1/2"
    D. 2"

19. _____ If a double- or triple-bowl sink is being installed, a(n) _____ fitting will be needed at the P-trap inlet.
    A. elbow
    B. tee
    C. sleeve
    D. check valve

20. The cold water supply tube is connected to the right-side angle valve, and the hot water supply tube is connected to the _____-side angle valve.

Name _____

21. _____ Shower drains are connected to the DWV piping during the _____.
   A. first rough
   B. second rough
   C. third rough
   D. last rough

22. _____ *True or False?* Installation of a shower valve is essentially the same as installation of a tub/shower valve.

23. _____ *True or False?* The shower valve has an outlet for the tub spout.

24. _____ The _____ assembly is installed by working through an access opening cut in the subfloor before the tub was set in place.
   A. drain
   B. overflow
   C. stopper
   D. All of the above.

25. _____ During the second rough, the tub/shower valve is secured to the frame of the building and connected to the _____ piping.
   A. hot water
   B. cold water
   C. DWV
   D. Both A and B.

26. _____ After installing the showerhead and tub spout, all connections should be checked for _____.
   A. plumb
   B. leaks
   C. excess putty
   D. exposed plumbing components

27. Whirlpool bathtubs are equipped with a _____ and controls to regulate the velocity of the water/air coming from the jets.
   A. special valve
   B. rotor
   C. switchboard
   D. compressor

28. _____ The two types of toilet installations are floor-mounted and _____.
   A. one-piece
   B. two-piece
   C. bidets
   D. wall-hung

29. _____ Most codes require a _____ at the toilet stub-out to allow shutting off the water in case of a malfunction.
   A. flange
   B. check valve
   C. valve
   D. cleanout

30. _____ The connection between a toilet and the DWV piping system is made with a _____ flange.
   A. pipe
   B. closet
   C. rubber
   D. plastic

31. _____ When installing a closet bowl, place the bowl temporarily over the flange to check for _____.
   A. watertight sealing
   B. airtight sealing
   C. levelness
   D. Both A and B.

32. Number the following steps (1–6) in the appropriate order for installing a toilet tank.
    A. _____ Place a spud washer over the water inlet hole of the bowl.
    B. _____ Attach the flush lever to the flush valve.
    C. _____ Install float rod and float ball on the flush valve, if necessary.
    D. _____ Carefully place the tank in position so the opening fits over the spud washer.
    E. _____ Secure the ballcock assembly to bottom of the tank.
    F. _____ Secure with tank bolts.

33. _____ *True or False?* Once the tank has been installed and secured, the plumber must check for leaks.

34. _____ When correcting the water level in the tank, the _____ on the ballcock is adjusted.
    A. float mechanism
    B. flexible water supply tube
    C. angle valve
    D. flush valve

35. _____ A bidet requires connecting the bowl outlet to the _____ piping system through a P-trap.
    A. hot water
    B. cold water
    C. DWV
    D. Both A and B.

36. _____ Since urinals are used frequently, they are usually fitted with _____.
    A. service sinks
    B. bidets
    C. flush valves
    D. pop-up stoppers

37. _____ _____ sinks are typically made of stainless steel or cast iron, and are usually located in a custodian's room.
    A. Commercial
    B. Service
    C. Self-rimming
    D. Residential

38. _____ Connecting the drain for water supply piping of a water fountain is essentially the same as connecting a _____ drain.
    A. floor
    B. lavatory
    C. sink
    D. bidet

39. _____ Many refrigerators are equipped with ice makers that require a _____ water supply tubing.
    A. 1/4″
    B. 1/2″
    C. 3/4″
    D. None of the above.

40. _____ A _____ valve can be used to connect the water supply pipe and an ice maker or a humidifier.
    A. shutoff
    B. check
    C. stop
    D. saddle

41. _____ Enclosed bath and shower stalls may be equipped with _____ that heat the water and dispense controlled amounts of steam into the stall.
    A. humidifiers
    B. dehumidifiers
    C. steam generators
    D. water heaters

Name _____ Date _____ Class _____

CHAPTER 25
Private Septic Systems

OBJECTIVE: You will be able to explain the operation of a simple septic system, list the essential materials and describe methods used in construction of a septic tank and leach field, and describe the construction and operation of alternative systems.

Carefully study the chapter and then answer the following questions.

1. _____ The Babylonians constructed the first known sewer system more than _____ years ago.
   A. 1,000
   B. 2,000
   C. 5,000
   D. 10,000

2. _____ _____ officials generally have jurisdiction over the installation of private waste disposal systems.
   A. Local plumbing
   B. Local health
   C. State health
   D. City inspector

3. _____ The two essential parts of a private waste disposal system are the _____ and the leach field.
   A. DWV piping
   B. distribution box
   C. septic tank
   D. well

4. Number the following steps in the appropriate order of the disposal process.
   A. _____ Solids settle to the bottom.
   B. _____ Liquid leaves the septic tank through a sealed pipe and enters the distribution box.
   C. _____ Waste enters the leach field through a perforated pipe.
   D. _____ Waste enters the septic tank.
   E. _____ Wastewater leaves the distribution box.
   F. _____ Water is absorbed into the ground.
   G. _____ Sealed pipe carries liquid waste to the leach field.

5. _____ *True or False?* Leach beds are commonly found in swampy areas or where flooding is common.

6. _____ A modern home produces about _____ gallons of waste daily for each person living in the house.
   A. 50
   B. 65
   C. 75
   D. 100

7. *True or False?* The rate at which the soil will absorb liquid is known as the percolation rate.

8. _____ Plantings over leach beds should generally be limited to _____.
   A. small trees
   B. grass
   C. small plants
   D. large plants

9. _____ The size of the _____ and the amount of gravel placed in the leach field runs affect the capacity of the system.
   A. pipes
   B. septic tank
   C. trenches
   D. All of the above.

10. _____ *True or False?* Water entering the ground from the leach field must *not* pollute wells.

11. _____ It is generally accepted that leach fields should be installed on ground at a(n) _____ the house.
    A. lower elevation than
    B. higher elevation than
    C. equal elevation to
    D. perpendicular angle to

12. _____ The _____ tank is probably the most common septic tank.
    A. steel
    B. fiberglass-reinforced
    C. concrete
    D. plastic

13. _____ The purpose of a _____ is to provide a place where the many lines of the leach field connect.
    A. septic tank
    B. central liquid zone
    C. baffle
    D. distribution box

14. _____ Many older leach fields were built using short lengths of _____ pipe laid end-to-end.
    A. smooth plastic
    B. corrugated plastic
    C. clay
    D. steel

15. _____ *True or False?* The useful life of a leach field is the most critical concern in designing a private septic system.

16. _____ A _____ valve is used with a dual leach field to direct effluent from the septic tank to one field while the other field "rests."
    A. flush
    B. diverter
    C. check
    D. stop

17. _____ _____ leach field systems consist of tunnels constructed of interconnected plastic units instead of pipe.
    A. Single
    B. Dual
    C. Chamber
    D. Aeration

18. _____ An aeration wastewater treatment plant injects _____ into the sewage, causing the growth of aerobic bacteria.
    A. air
    B. water
    C. bacteria
    D. chemicals

19. _____ Lagoon systems depend on _____ to digest waste.
    A. aerobic bacteria
    B. algae
    C. evaporation
    D. All of the above.

20. _____ *True or False?* The *Clivus Multrum* eliminates the need for flush water.

Name _____

21. _____ *True or False?* A closed system of waste treatment was developed primarily for homes where other types of systems could not be installed.

22. _____ A disposal method that distributes septic tank effluent through small pipes/tubes under pressure, where it is injected into the ground or sprayed over a special medium where bacterial action continues, is known as _____.
    A. composting
    B. aeration
    C. dosing
    D. filtration

23. _____ Strong chemicals used for cleaning drain lines should *not* be used with a septic tank because these chemicals kill _____ that are essential for it to function.
    A. bacteria
    B. gray water
    C. solids
    D. scum

24. _____ *True or False?* Minimizing the amount of chemicals that enter the septic system gives the bacteria ample time to break down the waste.

25. _____ Replacing toilets that require 3.5 or more gpf with toilets that require _____ gpf will help limit the amount of liquid entering the septic system.
    A. 3
    B. 2.5
    C. 1.6
    D. 1

26. _____ It is generally recommended that the tank be pumped at _____-year intervals.
    A. two
    B. three
    C. four
    D. five

27. _____ *True or False?* If the septic tank has a filter at the outlet, the filter should be cleaned regularly.

Notes

Name _____ Date _____ Class _____

CHAPTER 26
Storm Water and Sumps

OBJECTIVE: You will be able to explain the difference between storm sewers and sanitary sewers, install drainage piping that collects storm water runoff, and describe the function and installation of a sump pump.

Carefully study the chapter and then answer the following questions.

1. _____ The two separate drain/waste piping systems installed in a typical building are the sanitary sewer and the _____ piping system.
   A. DWV
   B. storm water
   C. water supply
   D. well

2. _____ Only water that occurs naturally on the property is collected by the _____ piping system.
   A. DWV
   B. sanitary sewer
   C. storm drainage
   D. water supply

3. _____ *True or False?* Most buildings have a combined DWV and storm drainage piping system.

4. _____ A _____ collects the groundwater that accumulates below the basement floor and outside the foundation walls.
   A. storm drain
   B. foundation drain
   C. downspout
   D. leader

5. _____ *True or False?* It is generally preferable to pipe storm water to a sump rather than to the surface.

6. _____ If a sump pump is installed, provisions must be made for backup _____.
   A. drain
   B. electrical power
   C. auxiliary drain
   D. None of the above.

7. _____ A foundation drain channels groundwater into a basin called a(n) _____.
   A. sump
   B. tank
   C. pipe
   D. evaporator

8. _____ A sump pump is turned on automatically by a(n) _____-controlled switch.
   A. automatically
   B. pneumatically
   C. float
   D. manually

9. _____ A _____ valve should be installed in the vertical pipe leading from the pump to prevent water in the pipe from flowing back into the sump when the pump shuts off.
   A. shutoff
   B. check
   C. float
   D. None of the above.

10. _____ The radioactive element _____ decomposes below the earth's surface and releases radon gas.
    A. radium
    B. carbon monoxide
    C. uranium
    D. radon

11. _____ If radon is a problem in the area, install a _____ sump to expel the radon gas to the outside of the building.
    A. sewage
    B. vented
    C. reducing
    D. one-pipe

12. _____ On lots with sufficient slope, it may be practical to omit the sump and _____.
    A. foundation drain
    B. floor drain
    C. sump pump
    D. roof drain

13. _____ *True or False?* Exterior stairwells, window wells, and ground or driveways that slope toward the building all pose storm drainage problems.

14. _____ A _____ drain should be installed at the bottom of a driveway that slopes toward the building.
    A. foundation
    B. storm
    C. roof
    D. trench

15. _____ *True or False?* The procedures for installing storm drainage piping are similar to those described for the installation of horizontal DWV piping below the concrete floor.

16. _____ _____ drains are installed alongside the footings inside the building.
    A. Foundation
    B. Hub
    C. Trench
    D. Floor

17. _____ *True or False?* Leaders from downspouts do *not* require a minimum slope.

18. _____ Trench drains use _____ to provide an area that will absorb water quickly.
    A. downspouts
    B. leaders
    C. gravel
    D. sand

Name _____ Date _____ Class _____

CHAPTER 27
Installing HVAC Systems

OBJECTIVE: You will be able to describe the basic operation of each HVAC system and its parts and explain the jobs plumbers are required to do while installing the systems.

Carefully study the chapter and then answer the following questions.

1. _____ A _____ provides moisture controls in the air-conditioning system.
   A. dehumidifier
   B. refrigeration unit
   C. humidifier
   D. Both A and C.

2. _____ Heat is transferred in _____ direction(s).
   A. one
   B. two
   C. three
   D. four

3. _____ *True or False?* Heat moves from a cooler object to a warmer object.

4. _____ *True or False?* Gas and electricity are the only two energy sources that produce heat.

5. _____ _____ have almost completely replaced individual area heaters.
   A. Central heating systems
   B. Floor furnaces
   C. In-wall heaters
   D. Fireplaces

6. _____ A furnace is composed of a burner, a heat exchanger, and a(n) _____.
   A. condenser
   B. fan
   C. heating coil
   D. air vent

7. _____ Combustion gases produced from the burner exit through a _____.
   A. duct
   B. fin
   C. flue
   D. pipe

8. _____ _____ perimeter heating systems pipe hot water to baseboard heating units and depend on convection currents to distribute the warm air.
   A. Forced-air
   B. Electric
   C. Gas-fired
   D. Hydronic

9. _____ In a hydronic heating system, water is heated by a(n) _____.
   A. furnace
   B. boiler
   C. electric baseboard
   D. None of the above.

Copyright Goodheart-Willcox Co., Inc.
May not be reproduced or posted to a publicly accessible website.

10. _____ A _____ valve releases excess water pressure in case a single-pipe, forced-circulation system malfunctions.
   A. gate
   B. relief
   C. T/P relief
   D. pressure-reducing

11. _____ A _____ valve limits the pressure of the incoming fresh water supply in a single-pipe, forced-circulation system.
   A. gate
   B. relief
   C. T/P relief
   D. pressure-reducing

12. _____ *True or False?* Air vents in a hydronic heating system allow any air that enters the system to escape at high points in the piping.

13. _____ *True or False?* A single-pipe hydronic system heats a building more uniformly than other systems.

14. _____ A multizone hydronic system can control the temperature of each room due to the installation of zone valves and _____ at each radiator.
   A. expansion tanks
   B. pressure-reducing valves
   C. thermostats
   D. air vents

15. _____ A _____ provides both hydronic heat and hot water for fixtures.
   A. multizone hydronic system
   B. single-pipe, forced-circulation system
   C. double-pipe, forced-circulation system
   D. combination hydronic boiler

16. _____ *True or False?* Geothermal, or ground-source, systems gain heat from the sun's heat energy.

17. _____ *True or False?* The efficiency of a geothermal system is nearly constant because the temperature of the earth or water remains nearly constant.

18. _____ The two basic cooling systems in common use are the remote system and the _____ system.
   A. integral
   B. condensing
   C. external
   D. manual

19. _____ In a cooling system, heat from the air is absorbed by a liquid refrigerant circulating inside the _____.
   A. condenser
   B. heat exchange coil
   C. evaporator
   D. starter

20. _____ A _____ valve is used with a heat pump to produce heated air on the heating cycle and cooled air on the cooling cycle.
   A. relief
   B. zone
   C. reversing
   D. reducing

21. _____ The amount of water vapor in the air compared to the amount it could actually hold at a given pressure and temperature is called the _____.
   A. air filtration rate
   B. relative humidity
   C. moisture level
   D. condensation rate

Name _____

22. _____ The comfort range for most people is _____ relative humidity.
   A. 20% to 60%
   B. 30% to 50%
   C. 30% to 70%
   D. 90% to 100%

23. _____ The process of changing air within an enclosed space by supplying and distributing fresh air and exhausting used air is called _____.
   A. evaporation
   B. ventilation
   C. heat exchange
   D. aeration

24. _____ *True or False?* Generally, the plumber's role in the installation of an HVAC system is limited to installation of fuel piping to the heating unit.

25. _____ Gas-fired heating units require piping of natural gas from the _____ to the heating unit.
   A. meter
   B. heat exchange coil
   C. evaporator
   D. None of the above.

26. _____ The purpose of a _____ is to catch any foreign particles in the gas piping that might damage the heating unit and valves.
   A. manual gas valve
   B. thermocouple
   C. sediment trap
   D. pilot filter

Match the burner unit parts with the proper names for questions 27–36.

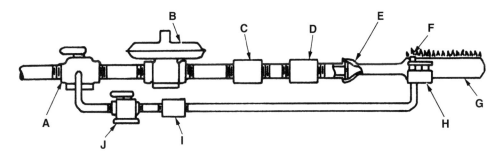

North American Heating and Air Conditioning Wholesalers Assoc.

27. _____ Safety shutoff valve

28. _____ Pilot burner

29. _____ Automatic main line valve

30. _____ B cock

31. _____ Burner

32. _____ Pressure regulator

33. _____ Venturi orifice

34. _____ Pilot filter

35. _____ Pilot generator

36. _____ A cock

37. _____ *True or False?* Solar heating units require installation of a storage tank for fuel oil.

38. _____ *True or False?* In forced-circulation hydronic systems, it is *not* necessary to pitch horizontal sections of piping.

39. _____ Drain valves in a hydronic system should be installed at the _____ points in the system.
    A. high
    B. low
    C. external
    D. vent

40. _____ _____ are installed to reduce vibration and prevent sagging of pipes.
    A. Drain valves
    B. Pipe hangers
    C. Tees
    D. Saddle valves

41. _____ A _____ is used to connect humidifiers to the water supply piping.
    A. drain valve
    B. pipe hanger
    C. tee
    D. saddle valve

Name _____ Date _____ Class _____

CHAPTER 28
Swimming Pools, Hot Tubs, and Spas

OBJECTIVE: You will be able to describe the components, the design considerations, and the installation of spas, hot tubs, and swimming pools.

Carefully study the chapter and then answer the following questions.

1. _____ *True or False?* The basic components of spas, hot tubs, and swimming pools are similar.

2. _____ _____ is a sand-concrete mixture that is applied by spraying the mixture over steel-reinforced material.
   A. Gravel
   B. Shotcrete
   C. Diatomaceous earth
   D. Fiberglass

3. _____ *True or False?* Hot tubs are usually constructed using wood.

4. _____ *True or False?* Many spas and hot tubs do *not* require direct plumbing and are commonly filled with a hose.

5. _____ The capacity of a hot tub or pool is measured in _____.
   A. quarts
   B. liters
   C. gallons
   D. pounds

6. _____ The _____ is the maximum number of people who will use the facility per hour.
   A. turnover rate
   B. capacity
   C. bathing load
   D. bathing limit

7. _____ The _____ is the frequency with which the total volume of water in the tub or pool is circulated through the filter.
   A. turnover rate
   B. capacity
   C. bathing load
   D. bathing limit

8. _____ Most public pools operate with a(n) _____ hour turnover rate so they can accommodate a larger number of people.
   A. 1–3
   B. 3–5
   C. 8–11
   D. 6–8

9. _____ _____ remove contaminants that would otherwise make the water unsuitable for bathing.
   A. Pumps
   B. Filters
   C. Skimmers
   D. Blowers

10. _____ _____ is the process of running water through the filter in the opposite direction.
    A. Filtering
    B. Backwashing
    C. Back siphoning
    D. Pumped filtration

Copyright Goodheart-Willcox Co., Inc.
May not be reproduced or posted to a publicly accessible website.

11. _____ _____ filters consist of a series of layers of porous material that is covered with diatomaceous earth.
    A. Sand
    B. DE
    C. Cartridge
    D. Backwash

12. _____ _____ are used to circulate the water through filters and the heater.
    A. Pumps
    B. Skimmers
    C. Blowers
    D. Hydrojets

13. _____ *True or False?* A spa outfitted with hydrojets requires a smaller pump.

14. _____ *True or False?* The design of a new installation of a large pool needs to be approved by the local health department.

15. _____ *True or False?* Spas and hot tubs usually do *not* have heaters.

16. _____ Floating debris is removed by _____.
    A. scum gutters
    B. sand filters
    C. skimmers
    D. Both A and C.

17. _____ _____ give spas and hot tubs the bubbling effect.
    A. Pumps
    B. Valves
    C. Pipes
    D. Blowers

18. _____ Swimming pools may have _____ to automatically add chlorine to the water for purification purposes.
    A. ozone generators
    B. filters
    C. ionizers
    D. chlorinators

19. _____ *True or False?* Hydrostatic pressure may cause damage to the piping when a pool is empty.

20. _____ A _____ is installed on the inlet side of the pump to remove hair and debris that would otherwise damage the pump.
    A. float valve
    B. filter
    C. strainer
    D. spout

Name _____ Date _____ Class _____

CHAPTER 29
Fire Safety and Irrigation Systems

OBJECTIVE: You will be able to describe the purpose and basic types of residential fire sprinkler systems. You will be able to list four basic considerations for operation of sprinkler irrigation systems, explain the importance of water pressure, and list factors that can cause water pressure loss. You will be able to describe the processes of designing and installing a lawn or garden sprinkler irrigation system and describe the differences between a sprinkler irrigation system and a drip irrigation system.

Carefully study the chapter and then answer the following questions.

1. _____ *True or False?* Residential fire sprinkler systems are less expensive than commercial systems.

2. _____ A _____ fire sprinkler system integrates cold-water pipes with fire sprinkler heads.
    A. multipurpose
    B. stand-alone
    C. quick-response
    D. None of the above.

3. _____ In a residential fire sprinkler system, the _____ control(s) when water is released.
    A. alarms
    B. piping
    C. sprinkler heads
    D. water meter

4. _____ *True or False?* Each of the four basic types of sprinkler heads operates in unique way.

5. _____ _____ sprinklers are designed to be installed in walls and generally have a deflector to direct water downward.
    A. Pendent
    B. Upright
    C. Concealed
    D. Sidewall

6. _____ The formula for computing flow rate is _____.
    A. $p=(q/k)^2$
    B. $k=qp^{0.5}$
    C. $q=kp^{0.5}$
    D. $q=(kp)^2$

7. _____ *True or False?* Design of residential fire sprinkler systems is typically done by engineers using specialized computer software.

8. _____ A _____ is an added safety feature that triggers an alarm when water begins to flow through the fire sprinkler system.
    A. fire alarm
    B. quick-response sprinkler head
    C. fire detector
    D. flow detector

9. _____ *True or False?* Lawn or garden sprinkler irrigation systems consist of an aboveground network of piping and sprinkler heads.

10. _____ Basic design considerations for a sprinkler irrigation system include providing controlled coverage, making use of existing _____, maintenance and repair costs, and being designed to prevent freeze damage.
    A. water piping
    B. water pressure
    C. sprinkler heads
    D. sprinkler systems

11. _____ The designer uses a _____ to determine the type and location of sprinkler heads on the property.
    A. foundation plan
    B. plot plan
    C. model
    D. formula

12. _____ _____ is the resistance to water flow exerted by the walls of the pipe and fittings.
    A. Flow resistance
    B. Friction
    C. Low pressure
    D. PSIC

13. _____ *True or False?* The designer can minimize pressure drop by using long pipe runs with numerous fittings.

14. _____ Drains should be placed at _____ points in the sprinkler irrigation system.
    A. low
    B. high
    C. equidistant
    D. no

15. _____ The three principal sprinkler head types are the spray, _____, and wave.
    A. upright
    B. pendent
    C. rotary
    D. fixed spray

16. _____ The _____-type sprinkler head is probably the most common type for residential lawn watering.
    A. pop-up, spray
    B. fixed spray
    C. pop-up, rotary
    D. wave

17. _____ The _____ spray head does *not* have a pop-up feature and is generally installed in flowerbeds.
    A. rotary
    B. fixed spray
    C. wave
    D. standard

18. _____ _____ sprinkler heads are designed to cover a large rectangular area.
    A. Fixed spray
    B. Pop-up action
    C. Rotary
    D. Wave

19. _____ *True or False?* All sprinkler irrigation systems are installed using exactly the same installation requirements and procedure.

20. _____ Working from the designer's plot plan, the installer will need to _____.
    A. locate the water source
    B. find a location for the controls
    C. lay out the sprinkler heads
    D. All of the above.

Name _____

21. _____ A(n) _____ cuts through the ground, forces the pipe into the cut, and buries the pipe with little damage done to the lawn.
    A. automatic pipe-laying machine
    B. trencher
    C. backhoe
    D. square-tipped shovel

22. _____ Polyethylene pipe is preferred over polyvinyl chloride pipe because of its resistance to _____ and water hammer.
    A. contraction
    B. decay
    C. expansion
    D. cold

23. _____ *True or False?* All pipes of a sprinkler irrigation system must slope uniformly toward one of the drain valves.

24. _____ *True or False?* Valve boxes control zone control valves, while the master control unit turns each section of the sprinkler system on and off automatically.

25. _____ A _____ system avoids much of the evaporation loss common to sprinkler irrigation systems that spray water into the air.
    A. drip irrigation
    B. flow-restricted
    C. rotary-type
    D. spray-type

26. _____ The flow of a drip irrigation system is controlled by _____ that emit 1/2 to 2 gallons of water per hour.
    A. zone control valves
    B. emitters
    C. filters
    D. hose bibbs

Match each of the parts of the drip irrigation connection with its correct name for questions 27–32.

27. _____ Tubing

28. _____ Filter

29. _____ Pressure regulator

30. _____ Faucet or hose

31. _____ Vacuum breaker

32. _____ Tubing adapter

Drip Irrigation Connection

Wade Mfg. Co.

Notes

Name _____ Date _____ Class _____

CHAPTER 30
Repairing DWV Systems

OBJECTIVE: You will be able to recognize DWV system problems, describe methods of checking and testing a plumbing system, and explain procedures for making proper plumbing repairs.

Carefully study the chapter and then answer the following questions.

1. _____ The three major groups of DWV problems are toilet (water closet), fixture, and _____ problems.
   A. sewer
   B. DWV piping
   C. water supply
   D. water pressure

2. _____ *True or False?* No repair should be made without first isolating the problem.

3. _____ The two common drainage malfunctions of a toilet are failure to drain and _____ near the base of the fixture.
   A. obstructions
   B. low water pressure
   C. leaks
   D. clogs

4. _____ The simplest way to clear a blocked toilet is to use a _____.
   A. water ram
   B. plunger
   C. closet auger
   D. cable

5. _____ *True or False?* A water ram uses air to apply high pressure to force the blockage on through the pipe.

6. _____ A _____ should be used if the force cup and water ram fail to remove the obstruction.
   A. closet auger
   B. plunger
   C. snake
   D. canister auger

7. _____ Before removing a toilet, turn off the water supply and flush the toilet to _____ the water from the tank.
   A. fill
   B. empty
   C. clean
   D. replace

8. _____ A _____ uses a motor-driven snake to remove obstructions.
   A. closet auger
   B. drain-cleaning machine
   C. dual-feed mechanism
   D. water ram

9. _____ When replacing a toilet, it is strongly recommended to install a new _____ to prevent a possible leak at the base of the bowl.
   A. closet flange
   B. closet bowl gasket
   C. float ball
   D. flush valve

Copyright Goodheart-Willcox Co., Inc.
May not be reproduced or posted to a publicly accessible website.

Identify the parts involved in removing a toilet for questions 10–18.

10. _____ Closet bolts
11. _____ Washer
12. _____ Wall flange
13. _____ Slip-joint nut
14. _____ Nut
15. _____ Cap
16. _____ Rubber washer
17. _____ Flexible supply
18. _____ Valve

Toilet

Goodheart-Willcox Publisher

19. _____ It is possible to look inside DWV piping for blockages using _____ equipment.
    A. drain-cleaning machine
    B. video inspection
    C. water jetter
    D. pipe locator

20. _____ *True or False?* A blocked stack should *never* be cleared by running a snake or drain-cleaning machine down the vent stack.

21. _____ *True or False?* If a snake catches or stops rotating while clearing blockage from a building drain, it is likely that the building sewer is broken.

22. _____ A cracked tank, loose closet bolts, and a(n) _____ are three causes of leaks near the base of a toilet bowl.
    A. damaged closet bowl gasket
    B. loose flush valve
    C. overfilled tank
    D. missing washer

23. _____ Five reasons for improper flow of a tub, lavatory, shower, sink, or bidet drain are blockage in the stopper mechanism, blockage in the fixture trap, blockage in the _____, blockage in the stack, and blockage in the building drain.
    A. water supply piping
    B. DWV branch piping
    C. sewer main
    D. vents

24. _____ A tub or lavatory is likely to have a _____ built into the fixture drain.
    A. spigot
    B. P-trap
    C. pop-up stopper
    D. closet flange

25. _____ If a P-trap is fitted with a _____, it is only necessary to remove the plug to inspect the condition of the trap.
    A. pop-up stopper
    B. cleanout
    C. slip-joint nut
    D. wall flange

108   Modern Plumbing Lab Workbook

Name _____

Match each of the parts of the P-trap assembly with its proper part name for questions 26–30.

26. _____ P-trap
27. _____ Union nut
28. _____ Wall flange
29. _____ DWV branch piping
30. _____ Slip joint nut

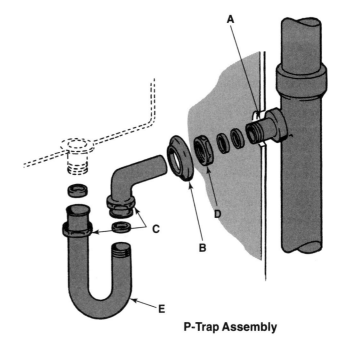

P-Trap Assembly

Goodheart-Willcox Publisher

Chapter 30 Repairing DWV Systems **109**

Notes

Name _____ Date _____ Class _____

CHAPTER 31
Repairing Water Supply Systems

OBJECTIVE: You will be able to recognize water supply problems, describe methods of checking and testing a water supply system, and explain procedures for making plumbing repairs.

Carefully study the chapter and then answer the following questions.

1. _____ Four major groups of problems within the water supply system are with toilets, faucets and valves, water supply piping, and _____.
   A. DWV piping
   B. water heaters
   C. sinks
   D. appliances

2. _____ Faucets are often a trouble spot and need repair due to _____.
   A. weak material
   B. heavy use
   C. water pressure
   D. lack of potable water

3. _____ A _____ faucet uses a rubberlike washer that is squeezed against a seat to shut off the flow of water.
   A. lavatory
   B. washerless
   C. compression
   D. disc

4. _____ A valve or faucet that does *not* completely stop water flow may be caused by a _____.
   A. worn or pitted seat
   B. deteriorated or broken washer
   C. defective handle or stem
   D. All of the above.

5. _____ If valve seats are pitted or worn, they must be _____.
   A. threaded
   B. refaced
   C. replaced
   D. Both A and B.

6. _____ Which of the following is *not* a possible cause for faucet or globe valve vibration and noise when the water is running?
   A. Loose washer.
   B. Deteriorated O-ring.
   C. Worn handle or stem.
   D. Worn faucet or valve base.

7. _____ *True or False?* The faucet handle should rotate without changing water flow.

Match each part of the faucet with its correct name for questions 8–19.

8. _____ Handle
9. _____ Packing
10. _____ Spline
11. _____ Bonnet
12. _____ Stem
13. _____ Spout
14. _____ Internal thread
15. _____ Faucet base
16. _____ Seat
17. _____ Brass screw
18. _____ Washer
19. _____ Stem threads

Faucet

Goodheart-Willcox Publisher

20. _____ Water leakage around the stem when the faucet or valve is opened is caused by deteriorated packing or a _____.
    A. worn valve seat
    B. worn stem
    C. broken closet bowl gasket
    D. All of the above.

21. _____ Most washerless faucets have a single _____ controlling both volume and temperature.
    A. spout
    B. valve
    C. handle
    D. cap

22. _____ The rotating ball faucet uses a _____ with ports to control water.
    A. handle
    B. ball
    C. cam housing
    D. shutoff valve

23. _____ *True or False?* When water flow from a spout is slow, the shutoff valve in the water supply piping should be checked first.

24. _____ A disc faucet uses a pair of _____ to control the water flow.
    A. escutcheon caps
    B. plastic discs
    C. ceramic discs
    D. aerators

25. _____ The only remedy for mineral deposits in piping is to _____.
    A. replace the affected pipes
    B. use a drain-cleaning machine
    C. use a snake
    D. replace the aerator unit

Name _____

Match each part of the rotating ball faucet with its correct name for questions 26–37.

26. _____ Cam housing

27. _____ Aerator

28. _____ Spout

29. _____ Cap

30. _____ Setscrew

31. _____ Ball assembly

32. _____ Cam rubber

33. _____ Diverter unit

34. _____ Stainless steel spring

35. _____ Seat washer

36. _____ O-ring washer

37. _____ O-ring

Rotating Ball Faucet

Goodheart-Willcox Publisher

38. _____ Water leaks around the joint between the spout and the faucet are caused by _____.
    A. deteriorated O-ring washers
    B. a worn spout
    C. a worn swing spout post
    D. All of the above.

39. _____ In a tank-type toilet, a faulty _____ will cause water to flow continuously.
    A. overflow tube
    B. float valve
    C. float ball
    D. All of the above.

40. _____ *True or False?* If the water in a toilet stops running when the float is lifted, the flush valve may be at fault.

41. _____ If a float ball does *not* rise to the surface, it is leaking and must be _____.
    A. repaired
    B. cleaned
    C. replaced
    D. All of the above.

42. _____ *True or False?* If the ball or flapper on a flush valve is worn, it will *not* seal properly.

43. _____ The two types of pressure flush valves are diaphragm and _____.
    A. piston
    B. flapper
    C. ball
    D. float

44. _____ In a diaphragm-type pressure flush valve, the _____ upsets the relief valve and causes water to flow past the diaphragm to the fixture.
    A. plunger
    B. vacuum breaker
    C. water stop
    D. closet trap

Chapter 31 Repairing Water Supply Systems 113

Match each part of the diaphragm-type pressure valve with its correct name for questions 45–53.

45. _____ Auxiliary valve
46. _____ Plunger
47. _____ Handle
48. _____ Lower chamber
49. _____ Upper chamber
50. _____ Diaphragm
51. _____ Bypass
52. _____ Inlet
53. _____ Outlet

Diaphragm-Type Pressure Valve

Sloan Valve Co.

54. _____ If there is no water in the toilet tank, and all the check valves in the supply line are open, the problem is in the _____.
    A. flush valve
    B. float valve
    C. water supply piping
    D. diaphragm pressure valve

55. _____ When water is leaking from the toilet tank, a _____ is a possible cause.
    A. cracked tank
    B. worn relief valve
    C. misaligned flush valve
    D. All of the above.

56. _____ *True or False?* The spud washer is located at the joint of the tank and bowl.

57. _____ Five problems that are likely to occur with the water supply piping are leaks at fittings, pipe leaks, broken pipes, flow restriction through pipe, and _____.
    A. contaminated water
    B. water hammer
    C. loose fittings
    D. increased flow

58. _____ *True or False?* Tightening one joint on threaded pipe does *not* loosen the joint on the other end of the pipe.

59. _____ Most plastic water supply piping is permanently _____ together.
    A. soldered
    B. brazed
    C. cemented
    D. welded

60. _____ Deterioration, impact/puncture, and _____ are the three principal causes of leaks in piping.
    A. heat damage
    B. freeze damage
    C. increased water flow
    D. blockage in the piping

61. _____ *True or False?* Installing a specially designed pipe clamp will only temporarily repair a punctured pipe.

Name _____

62. _____ Frozen metal pipe may be thawed by using _____, heat lamps, heat guns, and electric heaters.
   A. heat tape
   B. torches
   C. lighters
   D. your breath

63. _____ Plumbers must be sure the building water line is buried below the _____ line.
   A. sewer
   B. flow
   C. wall
   D. frost

64. _____ *True or False?* A few simple maintenance procedures can prolong the life of a water heater and increase operating safety.

65. _____ A water heater tank must be drained periodically to remove the _____ that collects at the bottom.
   A. sludge
   B. sediment
   C. effluent
   D. salt

66. Using the numbers 1 through 6, indicate the proper order of the steps to ignite the pilot on a water heater.
   A. _____ Release the reset button.
   B. _____ Depress and hold reset button.
   C. _____ Replace cover and turn thermostat to desired temperature.
   D. _____ Set thermostat valve on the pilot.
   E. _____ Hold reset button down for one minute or until thermocouple is heated.
   F. _____ Hold a lighted wooden match or charcoal lighter in front of pilot orifice.

67. _____ When an electric water heater does not produce hot water, check the _____ first.
   A. heating coil
   B. circuit breaker
   C. hot water piping
   D. thermostat

68. _____ *True or False?* The most obvious cause for shortage of hot water is a heater that is too large.

69. _____ If the water is too hot, the _____ is the most likely cause.
   A. thermocouple
   B. energy source
   C. thermostat
   D. dip tube

70. _____ *True or False?* An electric water heater should *not* be turned on before the tank is filled.

71. _____ A rumbling sound in a hot water heater tank as the water is heating is probably caused by _____ in the tank.
   A. air
   B. sediment
   C. cold water
   D. leakage

Notes

Name _____ Date _____ Class _____

CHAPTER 32
Remodeling

OBJECTIVE: You will be able to manage a plumbing remodeling job, replace fixtures and faucets without causing damage to other items, complete a roughed-in bathroom, relocate existing plumbing fixtures, determine if fixtures can be added to existing piping, and install new plumbing in an addition to an existing building.

Carefully study the chapter and then answer the following questions.

1. _____ In a remodeling job, existing piping refers to the piping that was installed during the _____-rough stage when the building was first constructed.
   A. first
   B. second
   C. third
   D. final

2. _____ The amount of modification to the DWV piping can greatly increase the time and _____ of a remodeling project.
   A. cost
   B. effort
   C. convenience
   D. All of the above.

3. _____ *True or False?* Remodeling work is most often done while the building is unoccupied.

4. _____ *True or False?* Code requirements must be considered during the planning of a remodeling project.

5. _____ *True or False?* The shortest route is always the best route for workers to get from the entrance to the work area.

6. _____ *True or False?* The plumber is responsible for communicating with other trades to establish responsibilities for the overall remodeling job.

7. _____ To protect the floor during remodeling, drop cloths, drop cloths covered with plywood cut to fit, or _____ can be used.
   A. non-slip rugs
   B. wide runners
   C. towels
   D. lumber channels

8. _____ To protect corners from damage during remodeling, fasten _____ on the corners.
   A. wood strips
   B. plywood channels
   C. wood caps
   D. plywood sheets

9. _____ Heating and air-conditioning return air ducts in the work area should be blocked to prevent _____ from entering.
   A. hot air
   B. cold air
   C. dust
   D. Both A and B.

10. _____ The first step when replacing a faucet is to _____.
    A. disconnect the water supply
    B. remove the P-trap
    C. turn off the water
    D. remove the nuts that secure the faucet

Copyright Goodheart-Willcox Co., Inc.
May not be reproduced or posted to a publicly accessible website.

11. _____ The second step when replacing a faucet is to _____.
    A. turn off the water
    B. remove the P-trap
    C. remove the nuts that secure the faucet
    D. disconnect the water supply

12. _____ If the P-trap is removed, the DWV stub-out should be plugged to prevent _____ from entering the room.
    A. propane gas
    B. sewer gas
    C. water condensation
    D. natural gas

13. _____ The first step in replacing a toilet is to _____.
    A. turn off water at the angle stop
    B. flush the toilet
    C. disconnect the water supply tube
    D. remove the closet bolt nuts

14. _____ *True or False?* Refinishing a tub is cheaper than replacing it.

15. _____ Removing an enameled cast-iron tub is difficult because of the fixture's _____ and size.
    A. drain
    B. piping
    C. material
    D. weight

16. _____ _____ are safety items that should be worn by a plumber when using a sledgehammer to break up an enameled cast-iron tub.
    A. Safety goggles
    B. Face shields
    C. Heavy gloves
    D. All of the above.

17. _____ *True or False?* The distance from the lavatory trap to the vent for the lavatory is typically limited by code.

18. _____ Developed length is measured along the _____ of the pipe and fittings.
    A. sides
    B. front
    C. end
    D. centerline

19. _____ When relocating fixtures, the first question that needs to be answered is, "Where are the _____ pipes located?"
    A. lavatory
    B. DWV
    C. gas
    D. water

20. _____ If the DWV piping is modified to change the location of a fixture, then the _____ piping will need to be changed also.
    A. gas
    B. water supply
    C. lavatory
    D. All of the above.

21. _____ When adding new fixtures, the _____ of the existing DWV piping must be evaluated to determine if it can carry the additional load.
    A. capacity
    B. number
    C. size
    D. pressure

Name _____

22. _____ Adding a lavatory to an existing bathroom simply requires adding _____ to an existing piping system.
    A. water supply fixture units
    B. drainage fixture units
    C. stacks
    D. P-traps

23. _____ The size of the branch hot and cold water pipes serving the bathroom is dependent on the _____.
    A. water supply fixture units
    B. developed length of the pipe and fittings
    C. water pressure in the piping system
    D. All of the above.

24. When new fixtures add more fixture units to the load than the existing DWV piping can carry, changes must be made to the _____.
    A. fixtures
    B. water supply piping
    C. DWV piping
    D. water pressure

Notes

Name _____ Date _____ Class _____

CHAPTER 33: Job Organization

OBJECTIVE: You will be able to describe how to become familiar with a new plumbing job, identify the materials and equipment necessary to complete a plumbing job, make better use of your time, describe the importance of teamwork on the job, and describe how to coordinate plumbing with the work of other trades.

Carefully study the chapter and then answer the following questions.

1. _____ For a new building, the _____ should be reviewed in order to understand the requirements of the job.
   A. drawings
   B. specifications
   C. plans
   D. All of the above.

2. _____ A(n) _____ is used to identify plumbing requirements and contains questions that can be answered by reviewing plans and specifications.
   A. detailed drawing
   B. checklist
   C. estimate
   D. plumbing drawing

3. _____ When a plumber needs to determine the size of pipe if plumbing drawings are not included for a residential structure, he or she should follow _____.
   A. architectural drawings
   B. building plan specifications
   C. federal code
   D. local code

4. _____ The building water and sewer lines are installed during the _____ stage.
   A. finish
   B. first-rough
   C. second-rough
   D. Either B or C.

5. _____ All pipe and fittings that will be covered in the finished structure are installed during the _____ stage.
   A. first-rough
   B. second-rough
   C. third-rough
   D. Either A or C.

6. _____ *True or False?* All materials to complete a plumbing job should be delivered at one time and dropped off in one place.

7. _____ The installation of fixtures, faucets, appliances, and water heaters is performed during the _____ stage.
   A. rough
   B. finish
   C. first-rough
   D. second-rough

8. _____ The _____ should indicate the materials, fixtures, and faucets to be used.
   A. specifications
   B. plumbing code
   C. mechanical code
   D. plumbing instructor

9. _____ *True or False?* When adding a bathroom to an existing building, the first-rough stage is eliminated.

10. _____ When preparing a list of plumbing materials and supplies, the quantity, size, and _____ of each item should be indicated.
    A. cost
    B. weight
    C. description
    D. All of the above.

11. _____ *True or False?* The list of materials should be divided into materials for the various stages of work.

12. _____ *True or False?* The number of people in the plumbing crew and their skills must be considered in planning.

13. _____ It is good practice to have slightly more materials and supplies available than is estimated to be installed in _____ of work.
    A. one day
    B. two days
    C. three days
    D. one week

14. _____ *True or False?* The plumber is responsible for ensuring that all tools are in safe and working condition.

15. _____ During construction, the most obvious need for coordination of the different trades relates to _____.
    A. payment
    B. scheduling
    C. equipment availability
    D. job responsibilities

16. _____ The keys to teamwork are to develop an environment that encourages teamwork, ensure that everyone knows what is to be done, _____, and reward people for achieving the desired outcome.
    A. share responsibilities
    B. make friends with coworkers
    C. dress appropriately
    D. plan ahead

17. _____ To develop an environment that encourages teamwork, employers need to emphasize the need for _____ among the members of the work group.
    A. responsibility
    B. discipline
    C. cooperation
    D. None of the above.

18. _____ Learning to work smarter primarily relates to _____.
    A. studying outside of work
    B. thinking ahead
    C. having a goal
    D. seeing the bigger picture

19. _____ Plumbers can be prepared to respond to injuries if they know how to contact the _____ and fire department.
    A. paramedics
    B. supervisor
    C. contractor
    D. All of the above.

20. _____ Generally, the _____ is expected to furnish all of the equipment necessary to perform an inspection.
    A. contractor
    B. plumber
    C. inspector
    D. supervisor

21. _____ *True or False?* To be a skilled plumber, it is necessary to achieve both quality and a reasonable level of speed in performing tasks.

Name _____ Date _____ Class _____

JOB 1 — Safety

TEXT REFERENCE: Study the appropriate section in Chapter 1, *Safety* before starting this job.

OBJECTIVE: After completing this job, the student will have demonstrated the ability to explain the importance of safety and inspect personal protective equipment, such as hard hats, safety goggles, safety shoes, hand tools, extension cords, and power tool cords to ensure that they are free from breaks or other damage.

Instructions

1. Explain the importance of safety and developing safe work habits.   Completed ☐

   _____

   _____

   _____

   _____

2. Inspect the safety items.   Completed ☐
   A. Hard hat
   B. Safety goggles
   C. Safety shoes
   D. Safety harness
   E. First-aid kit

3. Inspect the hand tools.   Completed ☐
   A. Wood chisel for sharpness
   B. Chisel for mushroom head
   C. Screwdriver for broken handle
   D. Fire extinguisher

4. Inspect the following items.   Completed ☐
   A. Extension cord
   B. Power tool cord
   C. Stepladder
   D. Rolling scaffold

Copyright Goodheart-Willcox Co., Inc.
May not be reproduced or posted to a publicly accessible website.

Notes

Name _____ Date _____ Class _____

JOB 2 — Math Calculations

TEXT REFERENCE: Study the appropriate section in Chapter 6, *Mathematics for Plumbers* before starting this job.

OBJECTIVE: After completing this job, the student will have demonstrated the ability to solve math problems using various math formulas and equations.

Instructions

Calculate and answer the following problems using the formulas discussed in the text material.

1. Solve the area of the following rectangles.   Completed ☐

   A. L = 25′, W = 15′   Area = _____
   B. L = 65′, W = 20′   Area = _____
   C. L = 19′, W = 6′   Area = _____
   D. L = 27.5′, W = 5.5′   Area = _____
   E. L = 17 yd, W = 9 yd   Area = _____
   F. L = 100′, W = 18′   Area = _____
   G. L = 40″, W = 16″   Area = _____
   H. L = 60 yd, W = 30 yd   Area = _____

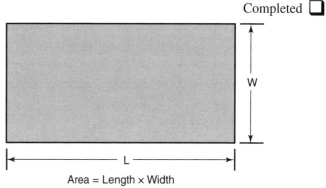

Area = Length × Width

Goodheart-Willcox Publisher

2. Solve the area of the following circles based on the information provided.   Completed ☐

   A. Diameter of 16′   Area = _____
   B. Radius of 8′   Area = _____
   C. Diameter of 12′   Area = _____
   D. Radius of 9′   Area = _____
   E. Diameter of 15′   Area = _____
   F. Diameter of 22′   Area = _____
   G. Radius of 6′   Area = _____
   H. Radius of 12′   Area = _____

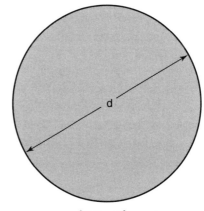

Area = πr^2
π = 3.14
r = Radius = ½ diameter

Goodheart-Willcox Publisher

Copyright Goodheart-Willcox Co., Inc.
May not be reproduced or posted to a publicly accessible website.

3. Solve the volume of the following rectangular tanks in cubic feet. Completed ☐
   A. L = 20′, W = 20′, H = 40′   Volume = _____
   B. L = 12′, W = 6′, H = 20′   Volume = _____
   C. L = 28′, W = 8′, H = 8′   Volume = _____
   D. L = 10′, W = 8′, H = 16′   Volume = _____
   E. L = 18′, W = 8′, H = 10′   Volume = _____

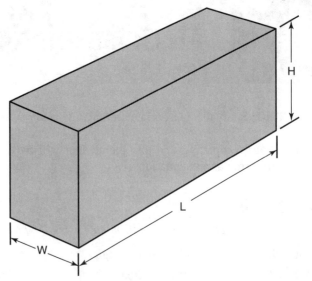

Volume = Length (L) × Width (W) × Height (H)

Goodheart-Willcox Publisher

4. Solve the volume of the following cylindrical tanks in gallons. Completed ☐
   A. Radius 5′, Height 15′   Volume = _____
   B. Radius 12′, Height 20′   Volume = _____
   C. Radius 3′, Height 12′   Volume = _____
   D. Diameter 10′, Height 30′ Volume = _____
   E. Diameter 20′, Height 60′ Volume = _____
   F. Radius 4′, Height 22′   Volume = _____

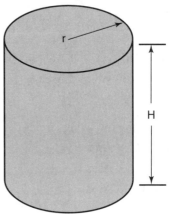

Volume = PI × radius squared × Height
This formula may be written: $V = \pi^2 h$

Goodheart-Willcox Publisher

Name _____ Date _____ Class _____

JOB 3: Calculating Slope on a DWV Horizontal Sewer

TEXT REFERENCE: Study the appropriate section in Chapter 5, *Leveling Instruments* before starting this job.

OBJECTIVE: After completing this job, the student will have demonstrated the ability to calculate the amount of slope on a horizontal sewer.

Instructions

1. Using the following illustration, calculate the slope of a horizontal sewer to be installed. Completed ☐

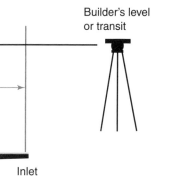

Total slope = Slope per foot × Feet of run

Goodheart-Willcox Publisher

---

2. Using the following information, calculate the total slope. Completed ☐

Calculating Slope			
Slope per Foot	×	Feet of Run	Total Slope
A. 1/8″		80 =	_____
B. 1/8″		100 =	_____
C. 1/4″		100 =	_____
D. 1/16″		100 =	_____
E. 1/2″		72 =	_____
F. 1/8″		48 =	_____
G. 1/2″		75 =	_____
H. 3/16″		100 =	_____

Goodheart-Willcox Publisher

Copyright Goodheart-Willcox Co., Inc.
May not be reproduced or posted to a publicly accessible website.

Notes

Name _____ Date _____ Class _____

JOB 4: Print Drawing

TEXT REFERENCE: Chapter 19, *Designing Plumbing Systems*

OBJECTIVE: After completing this job, the student will be able to design a drainage system for a simple bathroom and make an isometric drawing of this design.

Instructions

1. From the floor plan, design the drainage system for the bathroom. Completed

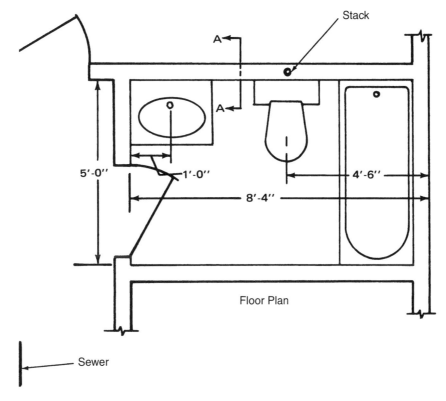

Floor Plan

Goodheart-Willcox Publisher

2. Make an isometric drawing of the design. Completed ☐

Name _____ Date _____ Class _____

JOB 5: Print Specification Interpretation

TEXT REFERENCES: Chapter 8, *Print Reading and Sketching*; Chapter 15, *Plumbing Fixtures*

OBJECTIVE: After completing this job, the student will have demonstrated the ability to read and interpret information provided on the fixture specification sheets.

Introduction

Print reading is a very important part of the plumbing installation process. Plumbers must be able to read the prints in order to locate walls and pipe chases during the first and second rough-in stages. It is imperative as an apprentice to begin learning how to read and interpret the information on the prints. In addition, plumbers must read and interpret (understand) plumbing fixture specification sheets in order to locate where to stub out drainpipes and water piping stub-outs for the various fixtures. This job will challenge the student to do just that.

Tools and Equipment

Lavatory Fixture Specification Sheet
Accessible Lavatory Specification Sheet
Round Front/Elongated Toilet Specification Sheet
Standard Bath Tub Specification Sheet

Instructions

1. Based on the information provided on the specification sheet for the *American Standard Declyn Wall-Hung Lavatory #0321.026,* answer the following questions.   Completed ☐

   A. What material was used to make the lavatory?

   _____

   B. The size of the waste at the wall is _____.

   C. The spread of the cold and hot water stub-outs is _____.

   D. The flood level rim for residential is _____ above finished floor.

   E. The flood level rim for ADA compliance is _____ above finished floor.

   F. The waste for residential is _____ above finished floor.

   G. If the flood level rim is 31″, what is the dimension to the top of the ledge back?

   _____

   H. The lavatory has a(n) _____ spread for mounting the faucet.

   I. The cold and hot water roughs out of the wall at _____ above finished floor.

   J. The center of the two openings for the lavatory wall bracket is installed how far above finished floor?

      a. Residential: _____

      b. ADA compliance: _____

2. Based on the information provided on the specification sheet for the *American Standard Accessible Lavatory #9141.011*, answer the following questions. Completed ☐

   A. What material was used to make the accessible lavatory?

   _____

   B. What is the overall dimension of the lavatory?

   _____

   C. The faucet opening has a(n) _____ spread.

   D. The flood level rim is _____ above the finished floor.

   E. The waste out at the wall is _____.

   F. The waste is located _____ above finished floor.

   G. What is the distance from the wall to the center of the waste outlet in the lavatory? _____

   H. The cold and hot water is roughed out at _____ above finished floor.

   I. The spread of the cold and hot water stub-outs is _____.

   J. The center of the faucet openings is _____ from the wall.

3. Based on the information provided on the specification sheet for the *American Standard Elongated Toilet #2399.012*, answer the following questions. Completed ☐

   A. What is the centerline of the outlet of the toilet from the wall? _____

   B. Based on the 12″ rough of the outlet, what is the distance from the back wall to the front of the toilet bowl? _____

   C. The toilet bowl is _____ wide.

   D. What is the dimension of the toilet tank lid? _____

   E. The top of the toilet bowl is _____ mm above the finished floor.

   F. The water supply rough-in is _____ above the finished floor.

   G. From the finished floor to the top of the tank including lid is _____.

   H. The water supply stub-out is _____ from the centerline of the DWV drain.

4. Based on the information provided on the specification sheet for the *American Standard Recess Bath #2391.202*, answer the following questions. Completed ☐

   A. The length of the tub is _____.

   B. The width of the tub is _____.

   C. The tub shown in the specification sheet is a _____ hand tub.

   D. What is the size of the drain outlet? _____

   E. The center of the drain outlet is _____ from the back (60″) side.

   F. The top of the front side of the tub is _____ above finished floor.

   G. The centerline of the valves is _____ above the finished floor.

   H. The centerline of the tub spout is _____ above the finished floor.

Name _____

American Standard

DECLYN™ WALL-HUNG LAVATORY
VITREOUS CHINA

DECLYN WALL-HUNG LAVATORY

- Vitreous china
- Rear overflow
- Soap depression
- Faucet ledge.
    Shown with 2000.101 Ceramix faucet (not included)

❏ **0321.026** With wall hanger (Illustrated)
Faucet holes on 102mm (4") centers

❏ **0321.075** For concealed arms support
Faucet holes on 102mm (4") centers

Nominal Dimensions:
483 x 432mm
(19" x 17")

Bowl sizes:
362mm (14-1/4") wide,
273mm (10-3/4") front to back,
152mm (6") deep

Fixture Dimensions conform to ANSI Standard A112.19.2

To Be Specified
❏ Color:
❏ Faucet*:
❏ Faucet Finish:
❏ Supplies:
❏ 1-1/4" Trap:
❏ Nipple:
❏ Concealed Arms Support (by others):

* See faucet section for additional models available

● Top of front rim mounted 864mm (34") maximum from finished floor.
MEETS THE AMERICAN DISABILITIES ACT GUIDELINES AND ANSI A117.1 REQUIREMENTS FOR PEOPLE WITH DISABILITIES

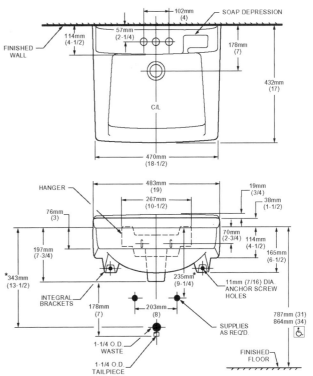

NOTES:
★ DIMENSIONS SHOWN FOR LOCATION OF SUPPLIES AND "P" TRAP ARE SUGGESTED.
PROVIDE SUITABLE REINFORCEMENT FOR ALL WALL SUPPORTS.
FITTINGS NOT INCLUDED AND MUST BE ORDERED SEPARATELY.
IMPORTANT: Dimensions of fixtures are nominal and may vary within the range of tolerances established by ANSI Standard A112.19.2.
These measurements are subject to change or cancellation. No responsibility is assumed for use of superseded or voided pages.

SPS 0321

LAV-053

© 1995 American Standard Inc.

Revised 4/97

American Standard Inc.

American Standard

 BARRIER FREE

ACCESSIBLE LAVATORY

ACCESSIBLE LAVATORY
VITREOUS CHINA

- Vitreous china
- Front overflow
- For concealed arms support (by others)
- Faucet ledge (faucet not included)

❏ **9141.011**
   Faucet holes on 102mm (4") centers
   (Illustrated)

❏ **9141.029**
   Faucet holes on 102mm (4") centers
   • Extra right-hand hole

❏ **9141.035**
   Faucet holes on 102mm (4") centers
   • Extra left-hand hole

❏ **9141.911**
   Faucet holes on 102mm (4") centers
   • Less overflow

❏ **9140.013**
   Faucet holes on 267mm (10-1/2") centers

❏ **9140.021**
   Faucet holes on 267mm (10-1/2") centers
   • Extra right-hand hole

❏ **9140.039**
   Faucet holes on 267mm (10-1/2") centers
   • Extra left-hand hole

❏ **9140.913**
   Faucet holes on 267mm (10-1/2") centers
   • Less overflow

❏ **9140.047**
   Center hole only

❏ **9140.947**
   Center hole only
   *Less overflow

NOTE: Roughing-in information shown on reverse side of page

Nominal Dimensions:
508 x 686m
(20" x 27")

Compliance Certifications -
Meets or Exceeds the Following Specifications:
• ASME A112.19.2 for Vitreous China Fixture

 Top of front rim mounted 864mm (34") from finished floor.
MEETS THE AMERICAN DISABILITIES ACT GUIDE-LINES AND ANSI A117.1 ACCESSIBLE AND USEABLE BUILDINGS AND FACILITIES - CHECK LOCAL CODES.

To Be Specified
❏ Color:
❏ Faucet*:
❏ Faucet Finish:
❏ Supplies with Stop:
❏ 1-1/4" Trap:
❏ Nipple:
❏ Concealed Arms Support (by others):
❏ Offset Grid Drain Assembly: 7723.018

* See faucet section for additional models available

CI-53

© 2004 American Standard Inc.

Revised 9/04
American Standard Inc.

Name _____

American Standard
♿ BARRIER FREE

ACCESSIBLE LAVATORY
VITREOUS CHINA

9141.011
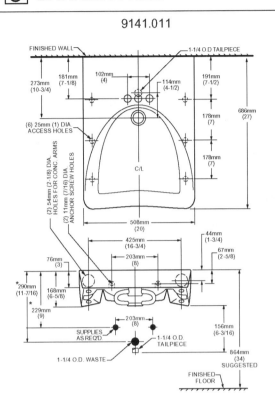

9141.029
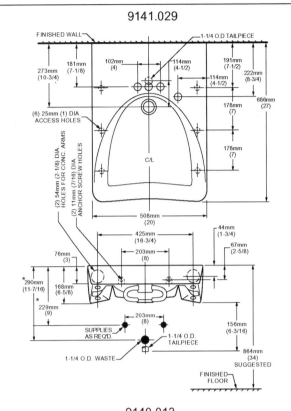

9141.035
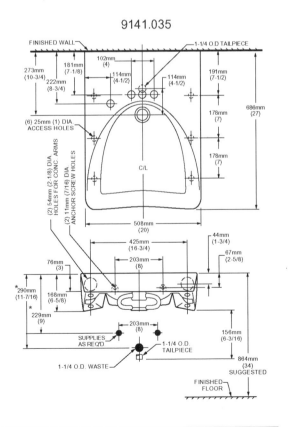

9140.013
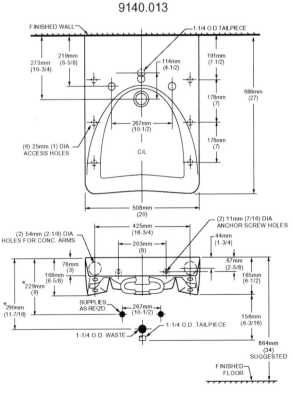

CI-54

© 2004 American Standard Inc.

American Standard Inc.

American Standard

COLONY™ ELONGATED TOILET
VITREOUS CHINA

COLONY™ ELONGATED TOILET

❏ **2399.012**
- Vitreous china
- Low-consumption (6.0 Lpf/1.6 gpf)
- Elongated siphon action jetted bowl
- Fully glazed 2" trapway
- American Standard Aquameter™ Water Control
- Large 10" x 8" water surface area
- Color-matched trip lever
- Sanitary bar on bowl
- 2 bolt caps
- 100% factory flush tested

❏ **3344.017** Bowl

❏ **4392.016** Tank

Nominal Dimensions:
762 x 508 x 743mm (30" x 20" x 29-1/4")

Fixture only, seat and supply by others

Alternate Tank Configurations Available:

❏ **4392.500** Tank complete w/Aquaguard Liner

❏ **4392.800** Tank complete w/Trip Lever Located on Right Side

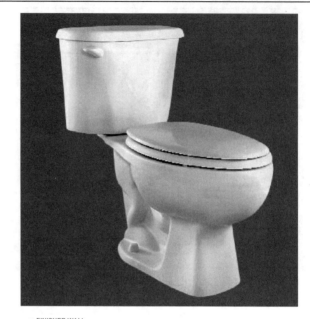

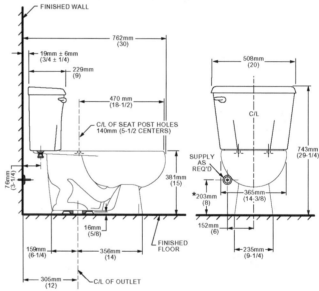

To Be Specified
- ❏ Color:
- ❏ Seat: American Standard #5324.019 "Rise and Shine" (with easy to clean lift-off hinge system) solid plastic closed front seat with cover. See pageTB-001.
- ❏ Seat: American Standard #5311.012 "Laurel" molded closed front seat with cover. See pageTB-001.
- ❏ Alternate Seat:
- ❏ Supply with stop:

NOTES:
THIS COMBINATION IS DESIGNED TO ROUGH-IN AT A MINIMUM DIMENSION OF 305MM (12") FROM FINISHED WALL TO C/L OF OUTLET.
* DIMENSION SHOWN FOR LOCATION OF SUPPLY IS SUGGESTED.
SUPPLY NOT INCLUDED WITH FIXTURE AND MUST BE ORDERED SEPARATELY.
IMPORTANT: Dimensions of fixtures are nominal and may vary within the range of tolerance established by ANSI Standard A112.19.2
These measurements are subject to change or cancellation. No responsibility is assumed for use of superseded or voided pages.

Compliance Certifications -
Meets or Exceeds the Following Specifications:
- ASME A112.19.2M (and 19.6M) for Vitreous China Fixtures - includes Flush Performance, Ball Pass Diameter, Trap Seal Depth and all Dimensions

SPS 2399.012

© 2000 American Standard Inc.

6/99

American Standard Inc.

Name _____

American Standard

PRINCETON™ RECESS BATH
AMERICAST® BRAND ENGINEERED MATERIAL

PRINCETON RECESS BATH
Americast® brand engineered material

- ❏ **2391.202** Right Hand Outlet
- ❏ **2391.202TC** Same as above w/tub cover
- ❏ **2390.202** Left Hand Outlet
- ❏ **2390.202TC** Same as above w/tub cover

- Acid resistant porcelain finish
- Recess bath with integral apron and tiling flange
- Integral lumbar support
- Beveled headrest
- Full slip-resistant coverage
- End drain outlet

PRINCETON RECESS BATH for Above Floor Rough Installation

- ❏ **2392.202** Left Hand Outlet for above floor installation
- ❏ **2392.202TC** Same as above w/tub cover
- ❏ **2393.202** Right Hand Outlet for above floor installation
- ❏ **2393.202TC** Same as above w/tub cover

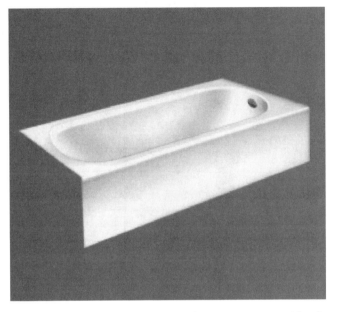

NOTE: Roughing-in dimensions shown on reverse side of page.

Nominal Dimensions: 1524 x 762 x 356mm (445mm for above floor installation)
60" x 30" x 14" (17-1/2" for above floor installation)

Bathing Well Dimensions: 1423 x 635 x 337mm (56" x 25" x 13-1/4")

Americast® brand engineered material is a composition of porcelain bonded to enameling grade metal, bonded to a patented structural composite.

Compliance Certifications -
Meets or Exceeds the Following Specifications:
- ASME A112.19.4 for Americast Plumbing Fixtures
- ASTM F-462 for Slip-resistant Bathing Facilities
- ANSI Z124.1 Ignition Test
- ASTM E162 for Flammability
- NFPA 258 for Smoke Density

To Be Specified
- ❏ Color:
- ❏ Bath Faucet*:
- ❏ Faucet Finish:
- ❏ Drain:
- ❏ Drain Finish:

* See faucet section for additional models available

Revised 5/98   © 2003 American Standard Inc.

BV-38

American Standard Inc.

American Standard

PRINCETON™ RECESS BATH
AMERICAST® BRAND ENGINEERED MATERIAL

NOTE: Roughing-in shown for 2390/2391 only. Refer to installation instruction for 2392/2393 above-floor installation.

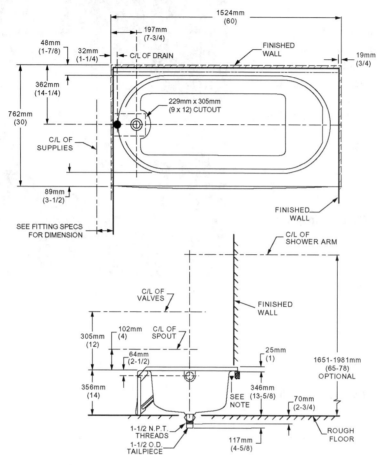

GENERAL SPECIFICATIONS FOR 2390/2391 BATHING POOL	
INSTALLED SIZE	60 x 30 x 14 In. (1524 x 762 x 356mm)
WEIGHT	110 Lbs. (50 Kg.)
WEIGHT w/WATER	460 Lbs. (209 Kg.)
GAL. TO OVERFLOW	42 Gal. (159 L)
BATHING WELL AT SUMP	42 x 19 In. (1067 x 483mm)
BATHING WELL AT RIM	56 x 25 In. (1423 x 635mm)
WATER DEPTH TO OVERFLOW	9-1/2 In. (241mm)
FLOOR LOADING (PROJECTED AREA)	37 Lbs./Sq.Ft. (175 Kgs./Sq.m)
PTS.	6.2
CUBE (FT³)	18.1

GENERAL SPECIFICATIONS FOR 2392/2393 BATHING POOL	
INSTALLED SIZE	60 x 30 x 17-1/2 In. (1524 x 762 x 445mm)
WEIGHT	119 Lbs. (54 Kg.)
WEIGHT w/WATER	469 Lbs. (213 Kg.)
GAL. TO OVERFLOW	42 Gal. (159 L)
BATHING WELL AT SUMP	42 x 19 In. (1067 x 483mm)
BATHING WELL AT RIM	56 x 25 In. (1423 x 635mm)
WATER DEPTH TO OVERFLOW	9-1/2 In. (241mm)
FLOOR LOADING (PROJECTED AREA)	38 Lbs./Sq.Ft. (182 Kgs./Sq.m)
PTS.	7.4
CUBE (FT³)	21.2

NOTES: LEFT HAND OUTLET SHOWN, RIGHT HAND REVERSE DIMENSIONS. (2391.202).
SHOWN WITH POP-UP C.D. & O.
REFER TO INSTALLATION INSTRUCTIONS SUPPLIED WITH FITTING.
CONCEALED PIPING NOT FURNISHED.
FITTINGS NOT INCLUDED AND MUST BE ORDERED SEPARATELY.
PROVIDE SUITABLE REINFORCEMENT FOR ALL WALL SUPPORTS.
▼ REFER TO INSTALLATION INSTRUCTIONS SUPPLIED WITH BATH.

IMPORTANT: Dimensions of fixtures are nominal and may vary within the range of tolerances established by ANSI Standard A112.19.4
These measurements are subject to change or cancellation. No responsibility is assumed for use of superseded or voided leaflet.

© 2003 American Standard Inc.

American Standard Inc.

Name _____ Date _____ Class _____

JOB 6: Fitting Identification

TEXT REFERENCE: Chapter 16, *Piping Materials and Fittings*

OBJECTIVE: After completing this job, the student will be able to identify the various types of fittings and joints.

Instructions

1. Identify the cast-iron hub fittings.   Completed ☐

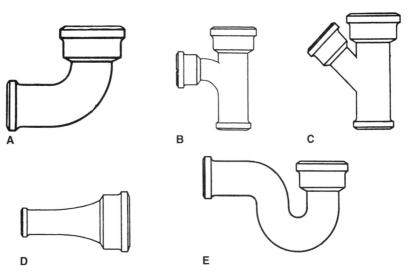

Cast Iron Soil Pipe Institute

A. _____ P-trap
B. _____ Y-branch
C. _____ Reducer
D. _____ 1/4 bend
E. _____ Sanitary tee

2. Identify the cast-iron no-hub fittings. Completed ☐

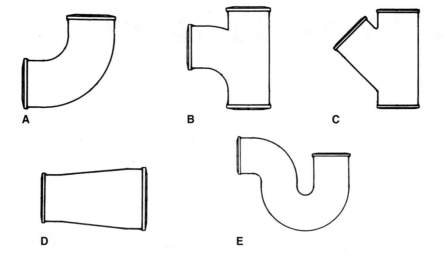

Cast Iron Soil Pipe Institute

A. _____ Y-branch
B. _____ Reducer
C. _____ P-trap
D. _____ Sanitary tee
E. _____ 1/4 bend

3. Identify the parts of the lead and oakum joint. Completed ☐

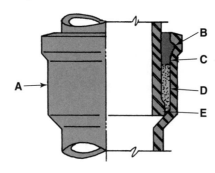

Cast Iron Soil Pipe Institute

A. _____ Hub
B. _____ Packed oakum
C. _____ Plain end/spigot bead
D. _____ 1″ deep lead
E. _____ Lead groove in hub

Name _____

4. Identify the parts of the compression joint. Completed ☐

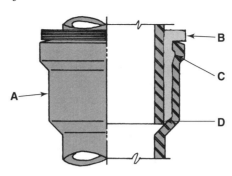

Cast Iron Soil Pipe Institute

A. _____ Gasket
B. _____ Hub
C. _____ Lead groove
D. _____ Plain end

5. Identify the parts of the no-hub joint. Completed ☐

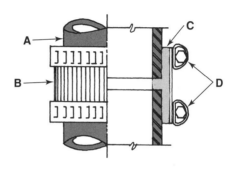

E.I. duPont de Nemours & Co.

A. _____ Stainless-steel shield
B. _____ Gasket
C. _____ Stainless-steel retainer clamps
D. _____ No-hub pipe

6. Identify the P-trap fittings. Completed ☐

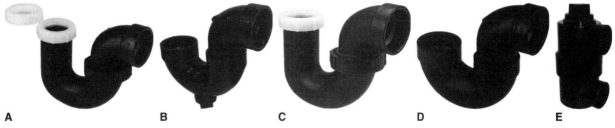

NIBCO, Inc.

A. _____ P-trap
B. _____ Swivel drum trap
C. _____ P-trap with slip joint
D. _____ P-trap with cleanout
E. _____ P-trap with union joint

Copyright Goodheart-Willcox Co., Inc.
May not be reproduced or posted to a publicly accessible website.

Job 6 Fitting Identification **141**

7. Identify the DWV plastic fittings. Completed ☐

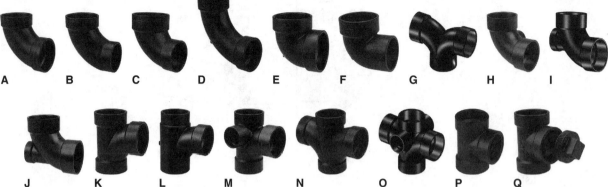

A. _____ 90° ell
B. _____ 90° closet ell
C. _____ 90° long turn ell
D. _____ 90° street ell
E. _____ 90° double ell
F. _____ 90° street vent ell
G. _____ 90° ell with high heel inlet
H. _____ 90° ell with side inlet
I. _____ 90° vent ell

J. _____ 90° ell low heel inlet
K. _____ Double sanitary tee
L. _____ Sanitary tee
M. _____ Sanitary street tee
N. _____ Sanitary tee with 90° side inlet
O. _____ Test tee
P. _____ Vent tee
Q. _____ Double sanitary tee/two 90° side inlets

8. Identify the copper pressure fittings. Completed ☐

A. _____ Union
B. _____ FIPT adapter
C. _____ MIPT adapter
D. _____ Flush bushing
E. _____ Drop tee
F. _____ Tee
G. _____ 45° fitting ell
H. _____ 90° ell
I. _____ Reducing coupling
J. _____ Coupling

Name _____

9. Identify the compression pressure fittings. Completed ☐

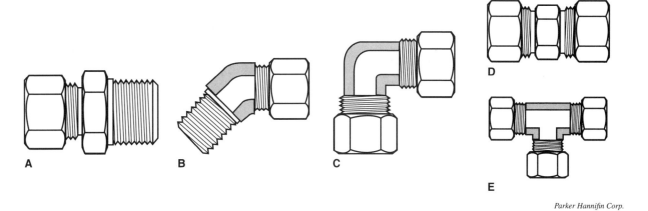

Parker Hannifin Corp.

A. _____ Compression union
B. _____ Compression ell
C. _____ Compression to MIPT adapter
D. _____ Compression tee
E. _____ Compression to MIPT 45° ell

10. Identify the malleable iron fittings. Completed ☐

A. _____ Cap
B. _____ Bushing
C. _____ Reducer coupling
D. _____ Tee
E. _____ 45° street ell
F. _____ Pressure fitting
G. _____ Drainage fitting

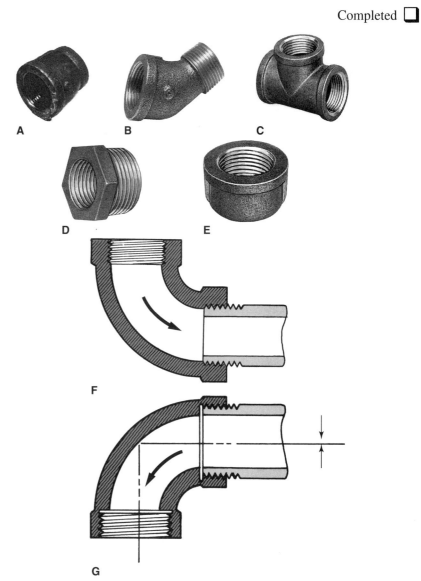

Goodheart-Willcox Publisher

Copyright Goodheart-Willcox Co., Inc.
May not be reproduced or posted to a publicly accessible website.

Job 6   Fitting Identification   **143**

11. Identify the four types of PEX fittings required to join PEX tubing. Completed ☐

    A. _____ Plug
    B. _____ Elbow
    C. _____ Tee
    D. _____ Coupling

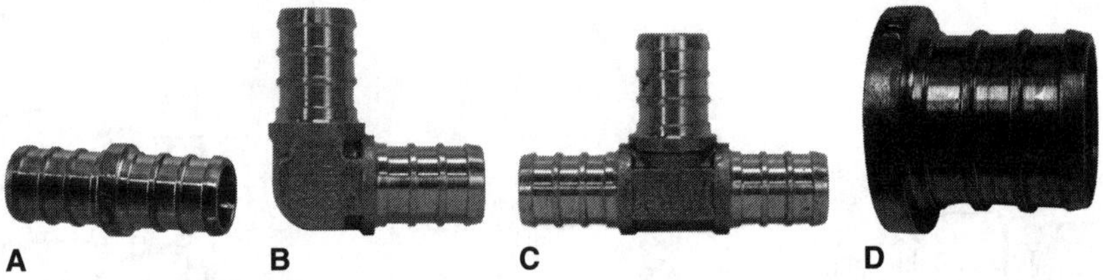

    A   B   C   D

    Goodheart-Willcox Publisher

Name _____ Date _____ Class _____

JOB 7 — Valve Identification

TEXT REFERENCE: Chapter 17, *Valves and Meters*

OBJECTIVE: After completing this job, the student will have demonstrated the ability to identify the parts of various valves.

Instructions

1. Identify the parts of the plastic ball valve. Completed ☐
   - A. _____ Handle
   - B. _____ Handle clip
   - C. _____ Stem
   - D. _____ Body
   - E. _____ Stem O-ring
   - F. _____ Ball seat
   - G. _____ Ball
   - H. _____ Carrier
   - I. _____ Carrier O-ring
   - J. _____ End connector
   - K. _____ Union nut
   - L. _____ Face seal O-ring

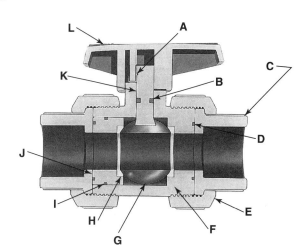

Celanese Plastics Co.

2. Identify the parts of the compression valve. Completed ☐
   - A. _____ Inlet
   - B. _____ Outlet
   - C. _____ Valve seat
   - D. _____ Washer
   - E. _____ Handle
   - F. _____ Packing nut
   - G. _____ Packing box
   - H. _____ Bonnet
   - I. _____ Stem
   - J. _____ Screw thread

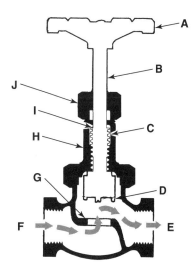

William Powell Co.

Copyright Goodheart-Willcox Co., Inc.
May not be reproduced or posted to a publicly accessible website.

145

3. Identify the parts of the diaphragm flush valve.
   A. _____ Handle
   B. _____ Auxiliary valve
   C. _____ Upper chamber
   D. _____ Diaphragm
   E. _____ Bypass
   F. _____ Outlet to fixture
   G. _____ Plunger
   H. _____ Inlet from water supply
   I. _____ Lower chamber

Completed ☐

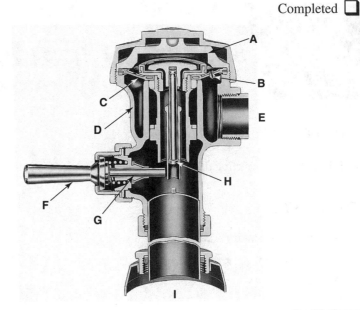

Sloan Valve Company

4. Identify the parts of a pressure regulator.
   A. _____ Pressure adjustment
   B. _____ Valve seat
   C. _____ Valve
   D. _____ Spring
   E. _____ Diaphragm

Completed ☐

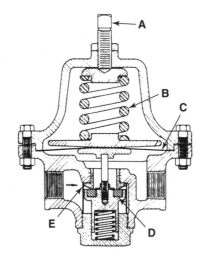

Watts Industries Inc.

5. Identify the three backflow prevention devices.
   A. _____ Dual-check valve with hose connection.
   B. _____ Dual-check valve for carbonated water.
   C. _____ Dual-check valve vacuum breaker.

Completed ☐

Photos courtesy of Watts

Name _____ Date _____ Class _____

JOB 8: DWV Pipe and Fitting Identification

TEXT REFERENCE: Chapter 16, *Piping Materials and Fittings*

OBJECTIVE: After completing this job, the student will be able to identify the fittings in a DWV pipe diagram.

Instructions

1. Identify the fittings in the DWV pipe diagram. Some letters may be used more than once.   Completed ☐

   A. _____ 2 × 2 × 2 wye
   B. _____ 2 × 2 × 1-1/2 san. tee
   C. _____ 2″ floor drain
   D. _____ 2″ long radius 90° ell
   E. _____ 1-1/2″ 45° ell
   F. _____ 2″ 45° ell
   G. _____ 2 × 1-1/2 × 1-1/2 san. tee
   H. _____ 2 × 1-1/2″ bushing
   I. _____ 2″ test tee with cleanout plug
   J. _____ 1-1/2″ trap adapter
   K. _____ 1-1/2″ 90° ell
   L. _____ 1-1/2″ P-trap

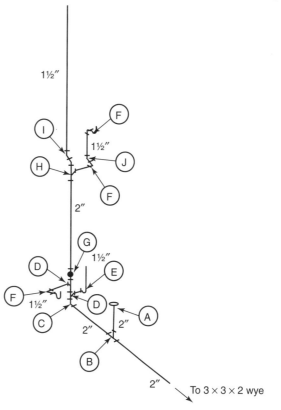

Goodheart-Willcox Publisher

2. Identify the fittings in the following DWV pipe diagram.

   A. _____ 3 × 3 × 3 test tee with cleanout
   B. _____ 3 × 3 long sweep sanitary ell
   C. _____ 4 × 3 closet flange
   D. _____ 3 × 3 sanitary tee with 2-1/2″ side inlet
   E. _____ 1-1/2 × 1-1/2 45° san. ell
   F. _____ 3 × 3 90° san. ell
   G. _____ 2 × 2 90° san. ell
   H. _____ 3 × 3 × 1-1/2″ tee
   I. _____ 1-1/2 × 1-1/2 90° ell
   J. _____ 1-1/4 × 1-1/4 90° san. ell
   K. _____ 1-1/4 × 1-1/4 45° san. ell

Completed ☐

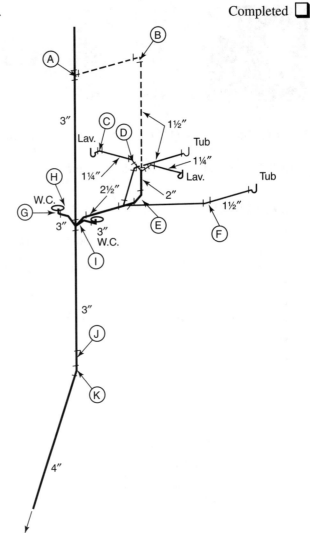

To building sewer

Goodheart-Willcox Publisher

JOB 9: DWV Fittings—No-Hub

TEXT REFERENCE: Chapter 21, *DWV Pipe and Fitting Installation*

OBJECTIVE: After completing this job, the student will be able to make a material list consisting of no-hub fittings and no-hub bands required for a residence consisting of two bathrooms, laundry room, and kitchen.

Instructions

Refer to the following isometric diagram to complete the job.

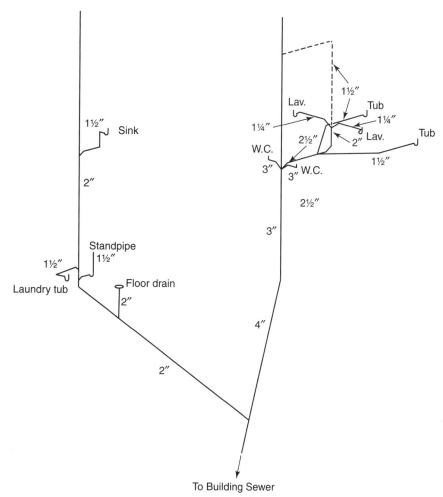

Goodheart-Willcox Publisher

1. Using the isometric diagram, list the no-hub fittings required for the system. Completed ☐
2. Using the isometric diagram, list the no-hub bands required for the system. Completed ☐

Quantity	No-Hub Fittings	No-Hub Bands	Size

Name _____ Date _____ Class _____

JOB 10: Soldering Copper Pipe and Fittings

TEXT REFERENCE: Chapter 11, *Soldering, Brazing, and Welding*; Chapter 16, *Piping Materials and Fittings*

OBJECTIVE: After completing this job, the student will have demonstrated the ability to join copper pipe and fittings using the soldering method. The connection must not leak when water pressure is applied. The instructor will determine the amount of pressure applied.

Introduction

Copper pipe and fittings may be joined by soldering. The connection is made watertight by using a filler metal known as solder. Heating the pipe and fittings with a plumber's torch melts the solder. The solder, which is now melted to a liquid state, is drawn into the socket of the fitting around the pipe and seals the connection. Although soldering is not difficult, the operation must be performed carefully for satisfactory results.

Tools and Equipment

Pipe vise
Tubing cutter
Measuring ruler or tape
Pencil for marking
Torch
Spark lighter
Solder (lead free)
Soldering flux/paste brush
Sandpaper (emery cloth)
Cleaning cloth
3′ of 3/4″ copper pipe
3/4″ 90° elbow
3/4″ copper male adapter
3/4″ copper cap
Safety goggles
Gloves

WARNING
Follow all safety rules and guidelines that may apply to performing this activity under the supervision of an instructor.

Instructions

1. Place the copper pipe in the pipe vise and secure. Completed ☐
2. Using a ruler, measure 12″ from the end of the pipe and mark with a pencil. Completed ☐
3. Cut the pipe with the tubing cutter. Completed ☐

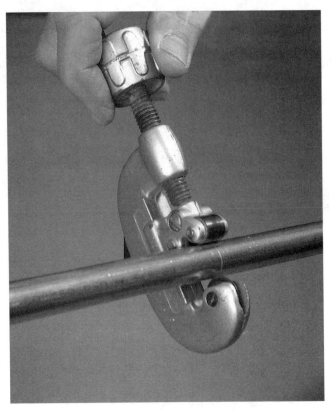

Goodheart-Willcox Publisher

4. Ream each end to remove the burr. Remove the burr from the inside of the tubing with a reamer. Completed ☐

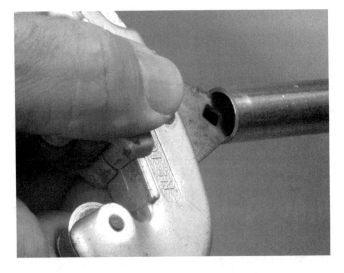

Goodheart-Willcox Publisher

5. Clean the outside end of the pipe in the vise and one end of the 12″ cut pipe. Clean the outside of the tubing with sandpaper cloth or steel wool. Completed ☐
6. Clean the inside of the sockets of the 90° elbow. Clean the inside of the fitting with a wire brush, steel wool, or sandpaper cloth. Completed ☐

Name _____

7. Immediately apply the proper amount of flux to all pipe and joint areas to be soldered. Apply flux to the outside of the tubing and the inside of the fitting. Completed ☐

Goodheart-Willcox Publisher

8. Assemble the fluxed pipes into the fitting. Push and turn to ensure that the pipes are bottomed against the inside shoulders of the elbow fitting. Completed ☐

9. Light the plumber's torch with the spark lighter. Completed ☐

WARNING
Do not use a cigarette lighter or a match to light the torch. This could cause a severe burn injury.

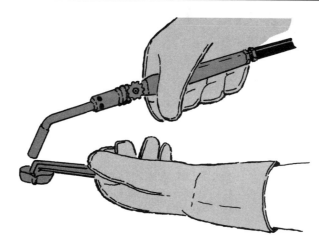

Goodheart-Willcox Publisher

10. Direct the heat onto the copper pipe and gradually move it toward the fitting. The inner cone of the flame should touch the metal. This is the hottest part of the flame. Completed ☐

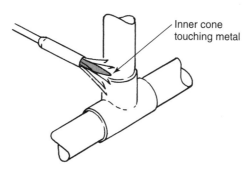

Inner cone touching metal

Goodheart-Willcox Publisher

Job 10 Soldering Copper Pipe and Fittings **153**

11. Slowly touch the end of the solder to the joint area when the flux starts to boil. Apply solder and feed solder into the joint until there is a silver ring all around the fitting. Completed ☐

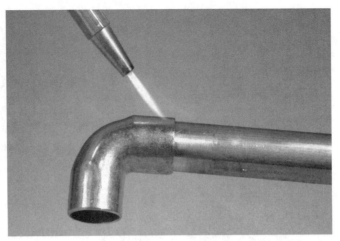

Goodheart-Willcox Publisher

12. Turn off the torch. Completed ☐
13. Wipe the joint with a clean, damp cloth while hot to remove excess solder and flux. Completed ☐

Goodheart-Willcox Publisher

14. Once the pipe has cooled, solder the cap on the 12″ pipe side of the elbow and the copper adapter on the 24″ pipe side of the elbow. Completed ☐
15. Under the supervision of the instructor, test the piping connections with water pressure to ensure that there are no leaks. Completed ☐
16. Clean up the work area and return all tools and material to the proper storage area. Completed ☐

Name _____ Date _____ Class _____

JOB 11: Joining PEX Piping Using Crimp Connections

TEXT REFERENCE: Chapter 22, *Installing Water Supply Piping*

OBJECTIVE: After completing this job, the student will be able to join PEX pipe and fittings using the crimp ring-type connection. The student will be evaluated on safety and selection of the correct tools and materials. The student will also be graded on following the procedural steps, returning tools to their proper place, and cleaning the work area.

Introduction

PEX pipe is used in pressure applications for building potable water systems. One method of joining PEX pipe is with crimp rings and crimp fittings. When properly joined, this process will provide a water tight connection necessary to provide safe operation for potable water piping.

Tools and Equipment

    1—Measuring tape
    1—Crimper tool
    1—Crimp ring gauge
    PEX pipe cutter
    3/4-inch PEX pipe
    3/4-inch PEX crimp rings

Instructions

1. Measure the pipe and mark for a desired length.   Completed ☐
2. Cut tubing square and remove burrs.   Completed ☐

NIBCO, Inc.

3. Slide the ring onto the tube.   Completed ☐

NIBCO, Inc.

4. Insert fitting into tubing all the way to the shoulder while positioning crimp ring 1/8″ to 1/4″ from end of tubing.   Completed ☐

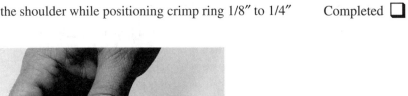

NIBCO, Inc.

5. Position jaws squarely over the crimp ring. Crimp the ring by completely closing the jaws of the crimping tool firmly.   Completed ☐

NIBCO, Inc.

6. Check the crimp ring with the gauge.   Completed ☐
7. Have the instructor check the connection.   Completed ☐

Name _____ Date _____ Class _____

JOB 12: Joining Cast-Iron No-Hub Pipe

TEXT REFERENCE: Chapter 21, *DWV Pipe and Fitting Installation*

OBJECTIVE: After completing this job, the student will be able to join no-hub cast-iron pipe to a no-hub cast-iron fitting using a stainless steel clamp with a neoprene sleeve. The student will be evaluated for safety and the proper selection of tools and materials. The student will also be evaluated on following the procedural steps, the return of tools to their proper place, and cleanliness of the work area.

Introduction

Cast-iron pipe is used primarily in nonpressure applications such as building drainage systems. Another term used for cast-iron pipe is *soil pipe*. Refer to the glossary in the textbook for further terminologies. The pipe and fittings are cast from gray iron. Cast-iron piping systems are selected because they are "quiet systems."

The purpose of this job is to join cast iron using the no-hub method, which allows for quicker installation. However, extra care must be taken to ensure proper support as specified by the Cast Iron Soil Pipe Institute. When properly installed, the no-hub system is trouble free.

Tools and Equipment

1—No-hub torque wrench
1—Stainless steel clamp with neoprene gasket
5′-2″ No-hub cast-iron soil pipe
2″ No-hub cast-iron 1/8 bend fitting (45° elbow)

> **NOTE**
> The following procedure is the same as joining no-hub cast-iron pipe end to end.

Instructions

1. Take the neoprene sleeve out of the stainless steel clamp.   Completed ☐

Goodheart-Willcox Publisher

2. Slide the stainless steel clamp onto the cast-iron soil pipe.   Completed ☐

3. Put the end of the pipe and fitting into the neoprene gasket. Completed ☐

Goodheart-Willcox Publisher

CAUTION
Be sure that the pipe and fitting fit to the center of the gasket since this will cause the fitting to be proper. Failure to perform step 3 correctly will cause the fitting to be loose (incorrect connection) and leak.

4. Slide the stainless steel clamp onto the neoprene gasket and tighten the clamp evenly with the torque wrench. Completed ☐

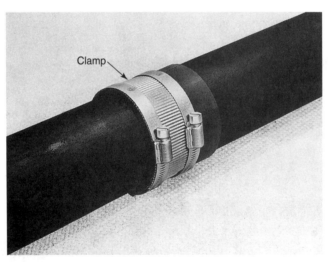

Goodheart-Willcox Publisher

5. Have the instructor check the connection to make sure the clamp is evenly tightened, square, and watertight at atmospheric pressure. Completed ☐

Name _____ Date _____ Class _____

JOB 13: Joining PVC Pipe

TEXT REFERENCE: Chapter 21, *DWV Pipe and Fitting Installation*

OBJECTIVE: After completing this job, the student will be able to join PVC pipe and fittings using the proper primer and solvent cement. The student will be evaluated on the safety and selection of the correct tools, materials, and solvent cement. The student will also be evaluated on the return of tools and materials, and the cleanliness of the work area.

Introduction

The use of polyvinyl chloride (PVC) pipe for plumbing is increasing rapidly due to low material cost and ease of installation. One advantage is that fewer hand tools are required. PVC is used in pressure (water supply) and non-pressure (DWV) applications. The technique of joining by solvent cement is relatively simple. When done correctly, it produces a leak-free connection. The student must always read the directions on the solvent cement container. The directions provide information regarding the hazards of using the cement and cleaner fluids.

> **NOTE**
> There is no such thing as "all-purpose cement." The strength of the cement is applicable to the pipe size. Therefore, refer to the manufacturer's specifications and directions. Remember, using the wrong strength cement can result in a leaking joint. Also, dissimilar plastics are not compatible to join together.

Tools and Equipment

1—6' plumber's rule
1—Fine tooth handsaw
1—Miter box
1—Half round file
1—Marking pencil
1—2" PVC pipe (longer than 24")
1—2" PVC 90° elbow
1—Wiping cloth
PVC primer and brush
PVC solvent cement and brush
Safety goggles

> **NOTE**
> Be sure the proper solvent is used, because there is no universal solvent cement.

Instructions

1. Using the 6′ plumber's rule, mark 24″ on the length of 2″ PVC pipe. Completed ☐
2. Place the pipe in the miter box and cut the pipe with the fine tooth saw. Be sure that the cut is square with no more than plus or minus 1/8″. Completed ☐

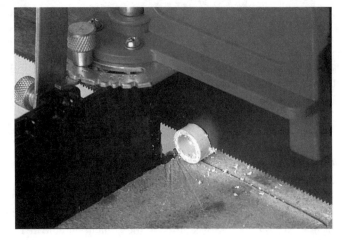

Goodheart-Willcox Publisher

3. Remove burrs from the inside of the pipe using the half round file. Completed ☐
4. Remove any ridges on the outer edges of pipe cut. Completed ☐
5. Use the wiping cloth to remove any dirt and grit from the pipe end and from the inside of the 2″ PVC elbow sockets. Completed ☐
6. Using the primer application brush, apply an even coat of primer to the end of the pipe as far as it will be inserted into the fitting. Apply an even coat of primer in the fitting socket. Completed ☐

> **WARNING**
> Take care while using the primer and solvent cement because the fumes from both are extremely dangerous and highly flammable. Wear safety goggles.

7. Wait 15 seconds, then apply an even coat of solvent cement to the end of the pipe to the depth that will be inserted into the fitting. Completed ☐
8. Apply an even coat of solvent cement to the inside surface of the fitting. Completed ☐

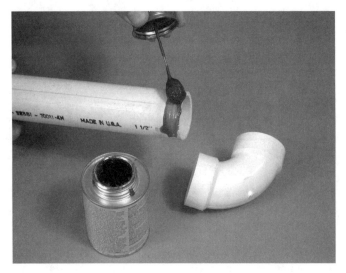

Goodheart-Willcox Publisher

Name _____

9. Immediately after applying the solvent cement, insert the pipe into the fitting, twisting the pipe one-quarter of a turn until it bottoms in the socket. Make the twist in one direction only. Hold the joint firmly in place for about 10 seconds (longer in cooler weather) to allow the two surfaces to start bonding together. Completed ☐

Goodheart-Willcox Publisher

10. Check the ring of solvent cement to see if it has been pushed out all the way around the joint during assembly and alignment. If the ring does not go all the way around the joint, you have not used enough solvent cement and the joint could leak. Completed ☐

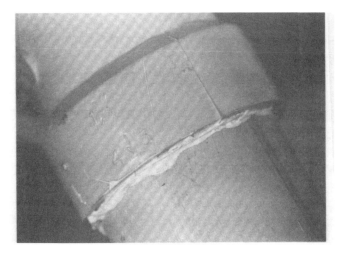

Goodheart-Willcox Publisher

11. Wipe off the excess solvent if the ring looks complete. Removing the excess solvent helps the joint cure faster. Completed ☐

12. Return tools and materials to their proper places and clean the work area. Completed ☐

13. Have the instructor check the joint for proper measurement, proper alignment, and leaks. Completed ☐

Notes

Name _____ Date _____ Class _____

Joining Galvanized Steel Pipe and Sleeve Coupling

TEXT REFERENCE: Chapter 21, *DWV Pipe and Fitting Installation*

OBJECTIVE: When provided with the proper tools and materials, the student will be able to join steel pipe with a sleeve coupling, producing a joint that will not leak when pressure is applied. The student will be evaluated by the instructor on the procedural steps of this job and will also be graded on the selection of tools, the return of the tools selected, and the cleanliness of the work area.

Introduction

Galvanized steel pipe is used for hot and cold water distribution, steam and hot water heating, gas and air piping, and drainage and vent piping. One method of joining galvanized steel pipe is by threading the pipe and then using it with threaded fittings. The threads are tapered to form a watertight joint when tightened securely.

Tools and Equipment

2—14″ pipe wrenches
1—Yoke or chain vise
1—3/4″ ID galvanized sleeve coupling
1—3/4″ galvanized cap
2—3/4″ × 6″ galvanized nipples
Pipe joint sealer or Teflon™ tape

> **NOTE**
> The following procedure will be the same whether the pipe used is galvanized, black steel, or wrought iron.

Instructions

1. Secure one 3/4″ × 6″ nipple in the yoke or chain vise.   Completed ☐

> **CAUTION**
> Do not tighten the vise on threads. This will damage the threads and cause the connection to leak.

Copyright Goodheart-Willcox Co., Inc.
May not be reproduced or posted to a publicly accessible website.

2. Apply pipe joint sealer or Teflon™ tape to protect the threads on the nipple in the vise. Completed ☐

Goodheart-Willcox Publisher

3. Screw the sleeve coupling onto the pipe threads and turn it clockwise until it is hand tight. Completed ☐
4. Place the wrench on the sleeve coupling and turn it clockwise one to three turns until firmly tight. Completed ☐

Goodheart-Willcox Publisher

> **CAUTION**
> Do not overtighten. Overtightening will spread the sleeve coupling and cause the joint to leak.

5. Apply pipe joint sealer or Teflon™ tape to one end of the other 3/4″ × 6″ nipple and turn clockwise in sleeve coupling until hand tight. Completed ☐
6. Place the pipe wrench on the second nipple and turn it clockwise one to three turns until firmly tight. Completed ☐

> **CAUTION**
> Do not overtighten because this will spread the sleeve coupling and cause the joint to leak.

7. Apply pipe joint sealer or Teflon™ tape to protect the threads on the nipple. Completed ☐
8. Screw the 3/4″ cap on threads of the nipple and turn clockwise until hand tight. Completed ☐
9. Place the pipe wrench on the cap and turn it clockwise one to three turns until firmly tight. Completed ☐
10. Have the instructor check for leaks by applying water pressure. Completed ☐

Name _____ Date _____ Class _____

JOB 15: Cutting and Threading Galvanized Steel Pipe

TEXT REFERENCE: Chapter 21, *DWV Pipe and Fitting Installation*

OBJECTIVE: When provided with the proper tools and materials, the student will be able to thread galvanized steel pipe using the ratchet threader. The student will be evaluated by the instructor on the procedural steps of this activity and will cut pipe within plus or minus 1/8″ of the measurement provided. The student will also be graded on the selection and return of tools and cleanliness of the work area.

Introduction

Galvanized steel pipe is used for hot and cold water distribution, steam and hot water heating, gas and air piping, and drainage and vent piping. One method of joining galvanized steel pipe is by threading the pipe, which may then be used with threaded fittings. The threads are tapered to form a watertight joint when tightened securely.

Tools and Equipment

    1—Yoke or chain vise
    1—6′ plumber's rule
    1—Marking pencil
    1—Steel pipe cutter
    1—Steel pipe reamer
    1—1/2″ ratchet pipe threader
    21′-1/2″ ID galvanized steel pipe
    1—Wire brush
    Oiler
    Cutting oil
    Safety goggles

Instructions

1. Secure the pipe in the yoke or chain-type vise.   Completed

Pipe Vise (Yoke)

Chain-Type Vise

The Ridge Tool Co.

2. Using a 6′ ruler, measure 12″ from the end of the galvanized steel pipe and mark it with the marking pencil. Completed ☐

3. Open the steel pipe cutters by turning the T handle counterclockwise until the cutter slips over the 1/2″ galvanized pipe. Then turn the T handle clockwise to tighten the pipe cutter around the pipe on the mark that will line up with the cutter wheel. Snug the cutter around the pipe and make one revolution to see if the cutter wheel is tracking properly. Completed ☐

> **WARNING**
> Wear safety goggles.

4. If the cutter is tracking properly, continue making revolutions around the pipe and tighten the cutter 1/4 turn for each revolution until the pipe separates. Completed ☐

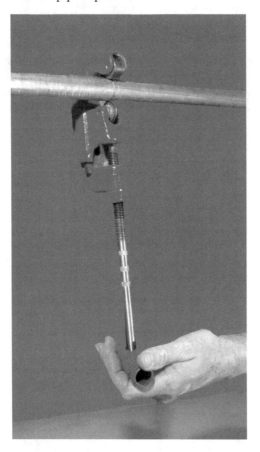

The Ridge Tool Co.

> **CAUTION**
> Do not overtighten. Overtightening will damage/break the cutter wheel or cause an excessive burr to form inside the pipe.

Name _____

5. Ream the inside of the pipe to remove the burr caused by the pipe cutter and restore the inside diameter to its original size. Completed ☐

Goodheart-Willcox Publisher

> **CAUTION**
> Do not over ream the pipe, because this will make the pipe end weak.

6. Using the 1/2″ ratchet die threader, press the die head against the pipe with one hand, and then turn the ratchet handle clockwise with the other hand. Apply pressure until the die teeth catch. Lubricate the dies with cutting oil as they feed onto the pipe. Completed ☐

7. When the dies are flush with the end of the pipe, the threads are at standard length. Completed ☐

> **CAUTION**
> Use plenty of cutting oil to preserve the pipe dies.

8. Remove die threaders by reversing the lock pin and turning the ratchet handle counterclockwise until the threaders are removed from the pipe. Completed ☐

9. Use a wire brush to clean the threads on the end of the pipe. Completed ☐

> **WARNING**
> Pipe threads are very sharp and can cut your hands and fingers if you are not careful.

10. Have the instructor check for proper threads and measurement of the pipe to plus or minus 1/8″. Completed ☐

11. Clean the work area and replace all tools and materials to their proper place of storage. Completed ☐

Notes

Name _____ Date _____ Class _____

JOB 16 — Water Supply

TEXT REFERENCE: Chapter 22, *Installing Water Supply Piping*

OBJECTIVE: After completing this job, the student will be able to make a copper or PVC/CPVC fitting material list for a water supply system.

Instructions

1. From the isometric diagram of the water supply system, make a copper or PVC fitting material list (take off list). Completed ☐

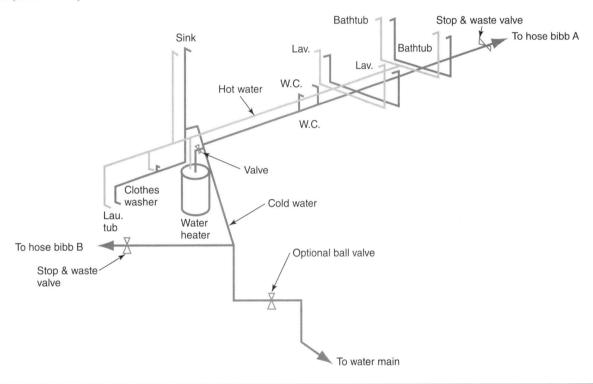

Quantity	Material	Size	Completed

Goodheart-Willcox Publisher

Notes

Name _____ Date _____ Class _____

JOB 17: Joining Copper Pipe Using Compression Fittings

TEXT REFERENCE: Chapter 24, *Installing Water Heaters, Fixtures, Faucets, and Appliances*

OBJECTIVE: After completing this job, the student will be able to join copper tubing using compression connectors. The connections will not leak when water pressure is applied.

Introduction
Copper pipe may be joined by compression fittings. The connection is made watertight by the compression of a brass ferrule around the copper pipe as the nut is tightened on the fitting.

Tools and Equipment
1—6″ crescent wrench
1—8″ crescent wrench
1—Tubing cutter
1—Ruler
1—Pencil or marking device
1—3/8″ OD compression type fitting
12″ × 3/8″ OD copper tubing

NOTE
Be sure the compression fitting has a brass ferrule.

Instructions

1. Using a ruler, measure 6″ from one end of the 3/8″ OD copper tubing and mark with a pencil.   Completed ☐
2. Cut the pipe on the mark and ream.   Completed ☐

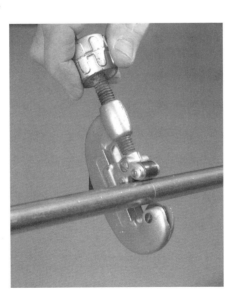

Goodheart-Willcox Publisher

3. Remove the compression nut and ferrule from one side of the compression fitting. Completed ☐

4. Slip the compression fitting nut over the end of the pipe, then slip the ferrule over the end of the pipe. Now insert the pipe end into the compression fitting socket. Completed ☐

Goodheart-Willcox Publisher

5. Slide the ferrule and nut up to the compression fitting and hand tighten. Completed ☐

6. Hold the compression fitting with the 8″ crescent wrench and tighten the compression nut with the 6″ crescent wrench until firmly tight. Completed ☐

7. Return tools and materials to their proper storage area. Clean the work area. Completed ☐

Name _____ Date _____ Class _____

JOB 18: Flaring and Connecting Copper Tubing

TEXT REFERENCE: Chapter 24, *Installing Water Heaters, Fixtures, Faucets, and Appliances*

OBJECTIVE: After completing this job, the student will be able to cut, ream, and flare 1/2″ copper tubing and make a flare connection that will not leak when pressure is applied.

Introduction

Copper pipe may be joined by using a flare type connection. This connection is made by flaring the tubing at a 45° angle, which fits to a beveled connector using a flare nut.

Tools and Equipment

1—Rigid #10 tubing cutters
1—6′ plumber's rule
1—Flaring tool
1—Marking pencil
1—Reaming tool
1—Roll of 1/2″ OD copper tubing
1—1/2″ OD flare nut
1—1/2″ MIP × 1/2″ flare adapter

Instructions

1. Using a 6′ plumber's rule, measure 8″ from the end of the roll of copper tubing and mark with the marking pencil. Completed ☐
2. Open the tubing cutter to fit over the tubing and place cutter wheel on the mark. Completed ☐
3. Tighten the handle on the cutter by turning it clockwise until it fits snug on the tubing. Completed ☐

Goodheart-Willcox Publisher

Copyright Goodheart-Willcox Co., Inc.
May not be reproduced or posted to a publicly accessible website.

4. Turn the tubing cutters around the pipe, tightening the handle 1/4 turn every three revolutions.   Completed ☐

CAUTION
Do not overtighten. Overtightening will damage the cutter wheel.

5. Turn and tighten repeatedly until tubing is cut and separated.   Completed ☐
6. Ream the burrs out of the tubing, using the reamer located in the tubing cutter.   Completed ☐

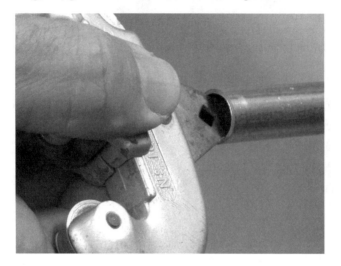

Goodheart-Willcox Publisher

CAUTION
Ream only down to the actual diameter of the pipe. Over reaming will cause the pipe walls to become thin and result in cracking and splitting while attempting to flare tubing.

7. Slide the flare nut over the pipe with the threads in the direction of the end of the pipe that is to be flared.   Completed ☐
8. Insert the tubing end into the flaring tool forming block that is the same as the pipe. Allow the end of the pipe to be flush with the flaring block.   Completed ☐
9. Tighten the blocks together.   Completed ☐

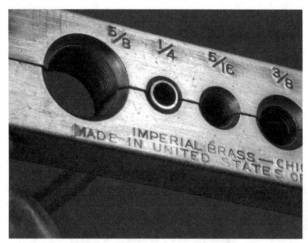

Imperial Eastman Corp.

174 Modern Plumbing Lab Workbook

Name _____

10. Slide the flare handle over the block until it centers over the tubing. Completed ☐
11. Lock the flare handle in place by slowly tightening the handle in the clockwise direction. Completed ☐

The Ridge Tool Co.

12. Tighten and flare the tubing until the handle becomes snug. Completed ☐

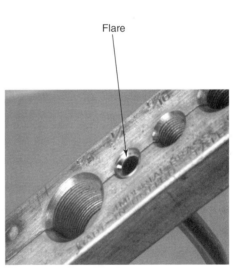

Goodheart-Willcox Publisher

CAUTION
Do not overtighten. Overtightening will crack and split the tubing.

13. Remove the flare tool by turning the handle counterclockwise. Completed ☐
14. Remove the flare block by loosening the blocks. Completed ☐
15. Check the flare for possible cracks or splits. Completed ☐
16. Slide the flare nut over the flare end of the tubing and tighten. Completed ☐
17. Check for leaks. Completed ☐
18. Return tools and equipment to their proper storage area. Clean the work area. Completed ☐

Notes

Name _____ Date _____ Class _____

JOB 19: Faucet and Rotating Ball Faucet Identification

TEXT REFERENCE: Chapter 31, *Repairing Water Supply Systems*

OBJECTIVE: After completing this job, the student will have demonstrated the ability to identify the parts of a compression faucet and rotating ball faucet.

Instructions

1. Identify the parts of the faucet.   Completed ☐
   - A. _____ Handle
   - B. _____ Packing
   - C. _____ Spline
   - D. _____ Bonnet
   - E. _____ Stem
   - F. _____ Spout
   - G. _____ Internal thread
   - H. _____ Faucet base
   - I. _____ Seat
   - J. _____ Brass screw
   - K. _____ Washer
   - L. _____ Stem threads

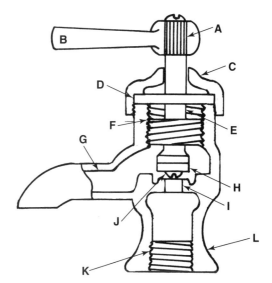

Goodheart-Willcox Publisher

2. Identify the parts of the faucet.   Completed ☐
   - A. _____ Cam housing
   - B. _____ Aerator
   - C. _____ Spout
   - D. _____ Cap
   - E. _____ Setscrew
   - F. _____ Ball assembly
   - G. _____ Cam rubber
   - H. _____ Diverter unit
   - I. _____ Stainless steel spring
   - J. _____ Seat washer
   - K. _____ O-ring washers
   - L. _____ O-ring

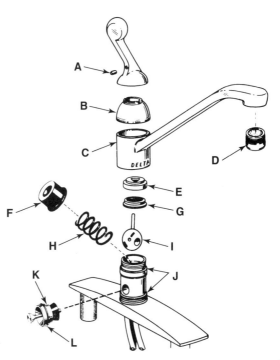

Goodheart-Willcox Publisher

Copyright Goodheart-Willcox Co., Inc.
May not be reproduced or posted to a publicly accessible website.

Notes

Name _____ Date _____ Class _____

JOB 20: Installing a Toilet (Tank-Type)

TEXT REFERENCE: Chapter 24, *Installing Water Heaters, Fixtures, Faucets, and Appliances*

OBJECTIVE: After completing this job, the student will have demonstrated the ability to install a tank-type toilet. The toilet will flush and function properly without any leaks.

Introduction

Toilets are either wall-hung or floor-mounted. The wall-hung units are often installed in commercial buildings to make floor cleaning easier. Toilets are available in a variety of styles, so it is necessary to refer to the manufacturer's instructions before beginning installation. The installation procedures discussed in this activity apply to most designs. Most codes require a valve at the toilet stub-out to allow shutting off the water supply to service the toilet. The tools and supplies required for installation of the stub-out valve depend on the type of pipe and fittings used for the cold water piping.

Tools and Equipment

8″ Crescent wrench
10″ Crescent wrench
Channel lock pliers
Tubing cutter
Level
Pipe joint compound
Toilet bowl wax ring
Angle shutoff valve, wall flange, toilet tank supply
Closet bolts, washers, nuts
Closet flange
Toilet unit (bowl and tank)
Toilet seat

Instructions

1. Turn off water supply serving the fixture branch. Completed ☐
2. Cut off the water stub-out. Completed ☐
3. Install the angle type shutoff valve. Completed ☐

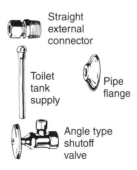

Chicago Specialty Mfg., Co.

4. Install the closet flange to the DWV piping system and flange bolts. Adjust the flange so the slots are equally centered from the back behind the toilet. Completed ☐

> **NOTE**
> The bottom of the flange should be flush with the finished floor.

5. To install the bowl, refer to the following exploded view and proceed as follows: Completed ☐

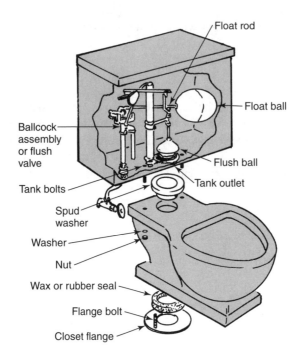

Goodheart-Willcox Publisher

   A. Place the bowl temporarily over the closet flange. Check for levelness. If it is not level, insert shims to level the bowl.

   B. Lift the bowl off the flange and turn it upside down. Fit the wax seal ring onto the discharge opening.

   C. Position the bowl carefully over the closet flange and closet bolts. Check again to make sure the bowl is level and squarely seated. Then, place washers on the closet bolts and install the closet nuts on the bolts. Tighten nuts securely and place bolt covers over bolts.

6. To install the tank, review the instructions provided with the toilet and again refer to the exploded view. The ballcock, flush valve, and flush valve lever may already be assembled. Completed ☐
   It is a good idea to check to make sure all connections are tight. To install the tank and fittings:

   A. Secure the ballcock assembly to the bottom of the tank, making sure the gasket provides a complete seal around the opening.

   B. Place a spud washer over the water inlet hole of the bowl.

   C. Carefully place the tank in position so that its opening fits over the spud washer. Press the tank into place.

   D. Secure with tank bolts.

   E. Install the float rod and float ball on flush valve if necessary.

   F. Attach the flush lever to the flush valve.

   G. Connect the water supply using a flexible water supply tube or flexible chrome plated copper toilet tank supply tube.

   H. Once the connections are secured, turn on the water supply, open the angle stop valve, and allow the tank to fill with water. Check all connections for leaks.

   I. Once any leaks have been corrected, flush the toilet to check for leaks at the toilet bowl seal.

   J. Install the toilet seat, making sure it is properly aligned over the bowl.

   K. Have the instructor check the installation.

Name _____ Date _____ Class _____

JOB 21 — Rough-In Residential Bathroom Group

TEXT REFERENCE: Chapter 20, *Preparing for Plumbing System Installation*; Chapter 21, *DWV Pipe and Fitting Installation*; Chapter 24, *Installing Water Heaters, Fixtures, Faucets, and Appliances*

OBJECTIVE: After completing this job, the student will have demonstrated the ability to rough-in the DWV piping system and the water supply piping system for a three-piece bathroom group.

Introduction

Roughing in a residential bathroom group provides an opportunity for students to demonstrate skills that have been learned in the plumbing class and lab. Basic plumbing installation skills will be used for installing a DWV piping system using PVC pipe and fittings and for installing a water piping system using copper pipe and fittings. Students will refer to the fixture manufacturer's specification instructions before beginning installation. Students will be required to rough-in the water supply and the drainage and vent system for a three-piece bath using PVC and copper pipe and fittings. The sole plate should be cut out for the stacks before installation is begun.

Tools and Equipment

- 4′ × 8′ bathroom mock-up with a 2″ × 6″ stud wall
- 6′ folding rule or steel tape
- Pencil or marking pen
- Copper tubing cutter and reamer
- PVC saw
- PVC reamer
- Battery powered drill—3/4″ or larger wood bit
- Claw hammer
- Keyhole saw
- Wood screws
- Standard screwdriver
- Soldering paste
- Copper cleaning brushes/emery cloth
- Channel lock pliers
- Soldering torch/tip
- Small gas tank cylinder
- Lead-free solder
- PVC glue and cleaner
- Hard hat
- Safety goggles
- Clean wiping rag
- Fire extinguisher
- Test plugs (3″, two—1-1/2″)
- Riser clamps (two—1-1/2″)

Instructions

1. Read the requirements for the DWV piping system and the water supply piping system before beginning the installation. Completed ☐

> **CAUTION**
> Comply with all safety rules and procedures.

2. Perform DWV system installation requirements. Completed ☐

   A. Soil and waste lines must be installed to a grade of not less than 1/8″ fall per foot. (Wooden blocks of 1/2″, 3/4″, and 1″ can be used to ensure proper grade.)

   B. Vertical pipes must be plumb.

   C. Horizontal vent lines must slope from the main stack back to the source.

   D. The 3″ vent stack must extend 12″ above the top plate to the center of the 3 × 2 vent tee.

   E. The top of the platform is considered finished floor unless directed otherwise.

   F. The centerline of fixtures shall match the floor plan and the manufacturer's specification sheets, plus or minus 1/4″.

   G. The closet flange shall be attached to the platform with wood screws.

   H. Riser clamps will be installed at the floor level of each stack.

   I. The DWV drainage system shall be water tested to check for any leaks.

3. Perform cold and hot water system installation requirements. Completed ☐

   A. The cold and hot water supply system shall be installed to comply with the isometric drawing shown on Drawing #3 and shall extend 8″ beyond the platform wall on the tub end and be tied back together as shown.

   B. Cold and hot water lines for the bathtub shall extend up to the height of the tub/shower valve and be looped together, as shown in Drawing #3.

   C. The water supplies for lavatory, toilet, and tub/shower shall be installed in accordance with the manufacturer's specification sheets, plus or minus 1/8″.

   D. Horizontal lines must be level, and vertical lines must be plumb.

   E. Fixture supply stub-outs shall extend a minimum of 5″ beyond the wall.

   F. The water supply system shall be water tested.

   G. Upon completion, clean the work area and return tools to storage.

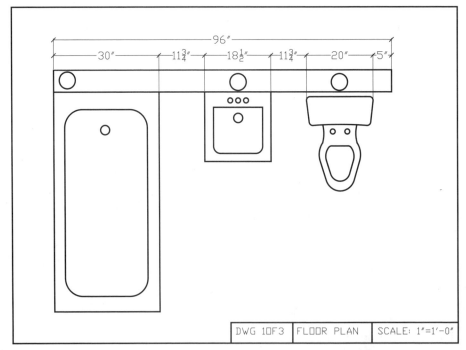

Goodheart-Willcox Publisher

Name _____

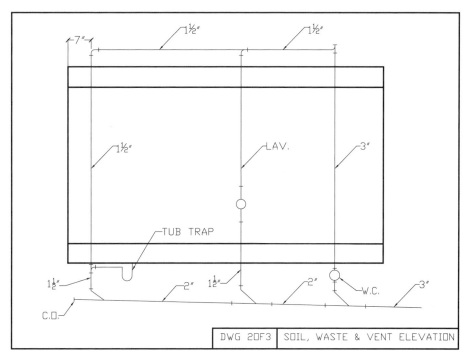

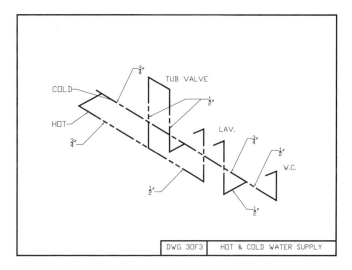

Job 21   Rough-In Residential Bathroom Group

American Standard

DECLYN™ WALL-HUNG LAVATORY
VITREOUS CHINA

DECLYN WALL-HUNG LAVATORY

- Vitreous china
- Rear overflow
- Soap depression
- Faucet ledge.
  Shown with 2000.101 Ceramix faucet (not included)

❏ **0321.026** With wall hanger (Illustrated)
Faucet holes on 102mm (4") centers

❏ **0321.075** For concealed arms support
Faucet holes on 102mm (4") centers

Nominal Dimensions:
483 x 432mm
(19" x 17")

Bowl sizes:
362mm (14-1/4") wide,
273mm (10-3/4") front to back,
152mm (6") deep

Fixture Dimensions conform to ANSI Standard A112.19.2

To Be Specified
❏ Color:
❏ Faucet*:
❏ Faucet Finish:
❏ Supplies:
❏ 1-1/4" Trap:
❏ Nipple:
❏ Concealed Arms Support (by others):

* See faucet section for additional models available

 ● Top of front rim mounted 864mm (34") maximum from finished floor.
MEETS THE AMERICAN DISABILITIES ACT GUIDELINES AND ANSI A117.1 REQUIREMENTS FOR PEOPLE WITH DISABILITIES

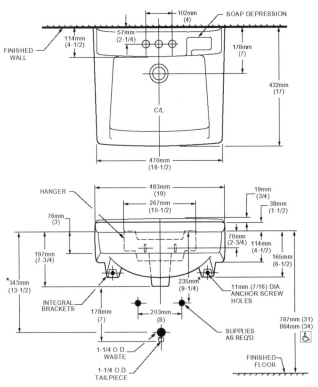

NOTES:
* DIMENSIONS SHOWN FOR LOCATION OF SUPPLIES AND "P" TRAP ARE SUGGESTED. PROVIDE SUITABLE REINFORCEMENT FOR ALL WALL SUPPORTS.
FITTINGS NOT INCLUDED AND MUST BE ORDERED SEPARATELY.
IMPORTANT: Dimensions of fixtures are nominal and may vary within the range of tolerances established by ANSI Standard A112.19.2.
These measurements are subject to change or cancellation. No responsibility is assumed for use of superseded or voided pages.

SPS 0321

LAV-053

© 1995 American Standard Inc.

Revised 4/97

American Standard Inc.

Name _____

American Standard

COLONY™ ELONGATED TOILET
VITREOUS CHINA

COLONY™ ELONGATED TOILET

❏ **2399.012**
- Vitreous china
- Low-consumption (6.0 Lpf/1.6 gpf)
- Elongated siphon action jetted bowl
- Fully glazed 2" trapway
- American Standard Aquameter™ Water Control
- Large 10" x 8" water surface area
- Color-matched trip lever
- Sanitary bar on bowl
- 2 bolt caps
- 100% factory flush tested

❏ **3344.017** Bowl

❏ **4392.016** Tank

Nominal Dimensions:
762 x 508 x 743mm (30" x 20" x 29-1/4")

Fixture only, seat and supply by others

Alternate Tank Configurations Available:

❏ **4392.500** Tank complete w/Aquaguard Liner

❏ **4392.800** Tank complete w/Trip Lever Located on Right Side

To Be Specified

❏ Color:

❏ Seat: American Standard #5324.019 "Rise and Shine" (with easy to clean lift-off hinge system) solid plastic closed front seat with cover. See pageTB-001.

❏ Seat: American Standard #5311.012 "Laurel" molded closed front seat with cover. See pageTB-001.

❏ Alternate Seat:

❏ Supply with stop:

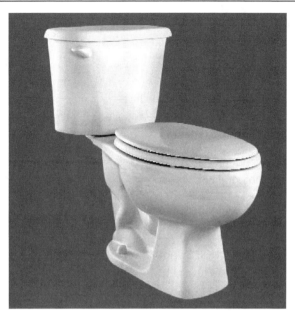

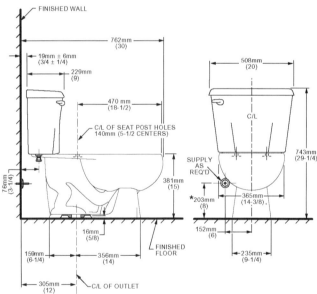

NOTES:
THIS COMBINATION IS DESIGNED TO ROUGH-IN AT A MINIMUM DIMENSION OF 305MM (12") FROM FINISHED WALL TO C/L OF OUTLET.
★ DIMENSION SHOWN FOR LOCATION OF SUPPLY IS SUGGESTED.
SUPPLY NOT INCLUDED WITH FIXTURE AND MUST BE ORDERED SEPARATELY.
IMPORTANT: Dimensions of fixtures are nominal and may vary within the range of tolerance established by ANSI Standard A112.19.2.
These measurements are subject to change or cancellation. No responsibility is assumed for use of superseded or voided pages.

**Compliance Certifications -
Meets or Exceeds the Following Specifications:**
- ASME A112.19.2M (and 19.6M) for Vitreous China Fixtures - includes Flush Performance, Ball Pass Diameter, Trap Seal Depth and all Dimensions

SPS 2399.012

© 2000 American Standard Inc.

American Standard Inc.

American Standard

PRINCETON™ RECESS BATH
AMERICAST® BRAND ENGINEERED MATERIAL

PRINCETON RECESS BATH
Americast® brand engineered material
- ❏ **2391.202** Right Hand Outlet
- ❏ **2391.202TC** Same as above w/tub cover
- ❏ **2390.202** Left Hand Outlet
- ❏ **2390.202TC** Same as above w/tub cover
  - Acid resistant porcelain finish
  - Recess bath with integral apron and tiling flange
  - Integral lumbar support
  - Beveled headrest
  - Full slip-resistant coverage
  - End drain outlet

PRINCETON RECESS BATH for Above Floor Rough Installation
- ❏ **2392.202** Left Hand Outlet for above floor installation
- ❏ **2392.202TC** Same as above w/tub cover
- ❏ **2393.202** Right Hand Outlet for above floor installation
- ❏ **2393.202TC** Same as above w/tub cover

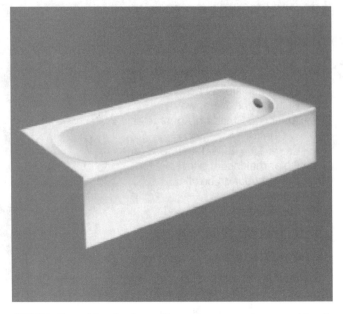

NOTE: Roughing-in dimensions shown on reverse side of page.

Nominal Dimensions: 1524 x 762 x 356mm (445mm for above floor installation)
60" x 30" x 14" (17-1/2" for above floor installation)

Bathing Well Dimensions: 1423 x 635 x 337mm (56" x 25" x 13-1/4")

Americast® brand engineered material is a composition of porcelain bonded to enameling grade metal, bonded to a patented structural composite.

Compliance Certifications -
Meets or Exceeds the Following Specifications:
- ASME A112.19.4 for Americast Plumbing Fixtures
- ASTM F-462 for Slip-resistant Bathing Facilities
- ANSI Z124.1 Ignition Test
- ASTM E162 for Flammability
- NFPA 258 for Smoke Density

To Be Specified
- ❏ Color:
- ❏ Bath Faucet*:
- ❏ Faucet Finish:
- ❏ Drain:
- ❏ Drain Finish:

* See faucet section for additional models available

BV-38

Revised 5/98   © 2003 American Standard Inc.

American Standard Inc.